Mathtastic Level 3 Numbers to 50 Teaching Book

Tracy Ashbridge

MEd, Grad Cert, PG Dip, BEd (Hons)

www.Mathtastic.com.au

Mathtastic

Learning Number Step by Step

Tracy Ashbridge

First Edition

2024 Brisbane

© Copyright 2024 Mathtastic: Tracy Ashbridge. All rights reserved

ISBN 978-0-6455822-2-2

The photocopiable pages, number problems booklet and word problems are available at:
http://mathtastic.com.au/mathtastic-resources/
Password: 3Mathtastic

Blue writing indicates this resource is available in the book.

Contents

How to use Mathtastic 7
Module 1 – Adding and Subtracting 0,1,2,3. 8
    Ideas for teacher lesson (1 hour) 8
    Reference cards Module 1 10
Module 2 – Add from largest, Subtract by counting back. 11
    Ideas for teacher lesson (1 hour) 11
    Reference cards Module 2 16
Module 3 – Rainbow Facts 17
    Ideas for teacher lesson (1 hour) 17
    Reference cards Module 3 20
Module 4 – Add and Subtract 10's. 21
    Ideas for teacher lesson (1 hour) 21
    Reference cards Module 4 23
Module 5 – Double and Halve 24
    Ideas for teacher lesson (1 hour) 24
    Reference cards Module 5 26
Module 6 – Near Doubles 27
    Ideas for teacher lesson (1 hour) 27
    Reference cards Module 6 30
Module 7 – Partition by Place Value 31
    Ideas for teacher lesson (1 hour) 31
    Reference cards Module 7 33
Module 8 – Add and Subtract by compensation. 34
    Ideas for teacher lesson (1 hour) 34
    Reference cards Module 8 38
Level 3 39
Numbers to 50 Resources 39
    Resources – not included 40
    Lesson Plan 41
        Homework - 15 mins per day 43
    Reference cards modules 1-8 45
    Module 1 Thinking Task - Newspaper 48
    MAB Subitizing cards 49
    Number Track 1-50 52

© Copyright 2024 Mathtastic: Tracy Ashbridge. All rights reserved

Number Tracks 1-50 partially completed. 53

Number Track 1-50 blank.. 56

Number line 1-50 57

Place Value Board 58

Number Square 1-50 completed. 59

Number Square 1-50 partially completed. 60

Number Square 1-50 blank... 61

Spinners for Magic Tens Game 62

    Easy 1-3 62

    Medium 1-6... 63

    Hard 4-9 64

Module 2: Subitizing with dominoes 65

Module 2 – Go fish, snap, memory or stealing game - same question in a different format. 69

Module 3: Blank Ten Frames 73

Module 3: Finger Subitizing Cards to 20 74

Module 3: Tally Marks to 20 78

One Hundred Square completed. 81

One Hundred Square partially completed. 82

One Hundred Square partially completed. 83

One Hundred Square blank... 84

Module 3: Go Fish/ Snap/ Memory/ Stealing game - Rainbow Facts to 50..................................... 85

Module 3: Follow on game. 90

Number line 1-100 92

Blank number line 94

.... 94

Module 4: Subitizing paddle pop sticks 95

Module 4 - Arrow cards 97

Module 4: Place Value Game – Three Bags Full... 101

Module 4 - Go Fish/ Snap/ Memory/ stealing game – place value. 103

Module 5: Ten frames to 50 – subitizing cards... 106

Module 5: Go fish, snap, memory, stealing game doubles. 109

Module 6: Number talks cards.. 115

Module 6: Rekenrek place value subitizing photos 117

Module 6: Abacus place value subitizing photos... 120

Module 6: Spike abacus 122

© Copyright 2024 Mathtastic: Tracy Ashbridge. All rights reserved

    Module 6: Go fish, memory, snap, stealing game - find a near double (and solve it)...................124

    Module 7 - Subitizing cards – all cards - photos 127

    Module 7: Round About game...154

    Module 7: Go fish, memory, snap, stealing game - find the same number partitioned in another way..156

Level 3 159

Numbers to 50 Workbook 159

Instructions 160

Setting out the student book 161

Coding the answers 163

Module 1 – Adding and Subtracting 0,1,2,3. 164

    Draw and solve..164

    Retrieval Practice – 10 mins per day..164

Module 2 – Add from largest, Subtract by counting back. 167

    Draw and solve..167

    Retrieval Practice – 10 mins per day..167

Module 3 – Rainbow Facts 169

    Draw and solve..169

    Retrieval Practice – 10 mins per day..169

Module 4 – Add and Subtract 10's 171

    Draw and solve..171

    Retrieval Practice – 10 mins per day..171

Module 5 – Doubles and Halves 173

    Draw and solve..173

    Retrieval Practice – 10 mins per day..173

Module 6 – Near Doubles 176

    Draw and solve..176

    Retrieval Practice – 10 mins per day..176

Module 7 – partition by place value 178

    Draw and solve..178

    Retrieval Practice – 10 mins per day..178

Module 8 – Add and Subtract by Compensation. 180

    Draw and solve..180

    Retrieval Practice – 10 mins per day..180

Level 3 182

Numbers to 50 182
Problem Solving book 182
   Recording Page 184
Module 1 - Add and Subtract 0,1,2,3 185
Module 2 - Add from largest, subtract by counting back 191
Module 3 - Rainbow Facts 197
Module 4 - Add and Subtract 10 203
Module 5 - Doubles 209
Module 6- Near Doubles 215
Module 7 – Partition by place value 221
Module 8 – Add and Subtract by compensation 227

How to use Mathtastic

Mathtastic can be used to teach 1:1 or small groups. The different level programs spiral through different levels of numbers, each level addressing the 8 number sense strategies:

1. Add 0,1,2, 3, Subtract 0,1,2, 3
2. Add from largest number by counting on, Subtract by counting back
3. Rainbow facts
4. Adding tens, subtract tens
5. Doubles/ halving
6. Near doubles
7. Partitioning numbers by place value
8. Adding and subtracting by compensating, Bridge 10 with a 9, Bridge 10 with 7 or 8, Round and adjust

There are 8 sections to each lesson:

1. Thinking problems – these are designed to be open ended and challenge the student to think mathematically. This is an opportunity for mathematical discussions and exploration.
2. Subitizing (sub rhymes with cube) – this is the skills of recognising a set of objects without counting and is a key skill which is not always established in students with difficulties in maths.
3. Counting patterns and objects – students need to develop a sense of the number line. This is a skill that students with difficulties are often weak in.
4. Number sense – each session there is a different focus working through the 8 areas of number sense. These are explained and modelled before applying the number sense concept to problems.
5. Game – many students with math difficulties can get anxious about maths and practising the skills through games is a less threatening way to gain the repetition they need. The games have been chosen to specifically practice the skill in focus and also allow for reasoning skills.
6. Word problems – students need to apply their knowledge in problems. These are organised by the 11 different ways of presenting addition and subtraction problems so students don't just learn to solve for the final answer but can be flexible to work around the problem.
7. Number problems – each session there are number problems related to the focus area and spaced retrieval of previous focus areas.
8. More games

Each module can be used as a lesson or can be split over several lessons depending on the time you have available and the speed the student works through the number sense strategies. The games can be repeated easily, and many have options to extend them.

Module 1 – Adding and Subtracting 0,1,2,3.
Ideas for teacher lesson (1 hour)

Thinking Problems	**Thinking task** You will need a page from the newspaper suitable for your students age (or print a page from an online news article). Use a highlighter to highlight all the numbers they can see in digits and in words. Can they read the numbers they have found? Are there any numbers repeated? Are there any patterns in the numbers? Example in the resources section.
Subitizing	**Subitizing** Use the MAB subitizing cards. The single digits are represented as ten frames presented pairwise. There are a mix of numbers from 20-50 and all the teen numbers as these are more difficult to master, mainly due to the inconsistencies in language. Can the students quickly recognise the number by how many tens and how many ones? You may need to break down the task into how many tens? How many ones? How many is that number?
Counting	**Counting – patterns and objects** 1. Use a number track to point and count to 50 forwards and backwards. Listen for careful articulation of 15 and 50, 13 and 30 etc. An extension of this would be to cut up the number track into chunks or individual numbers and for the student to order them. 2. Students to count and write the numbers to 50 using a partially completed and then a blank number track. Do this for counting both forwards and backwards. Monitor student number formation.
Number Sense	**Number sense focus area explain, explicit teach and model Examples and nonexamples** 1. In Levels 1 and 2 students have used number tracks for counting. As the student progresses, they move to number lines. This is a more abstract presentation. Place the number track and number line side by side and look for similarities and differences. Can the student find the same number on each line? 2. Use the number line or number track (depending on student understanding (if unsure go back to the number track) to explore the connection between addition and subtraction. For example, if I do 3+2, I get 5. However, if I reverse and do 5-2 you get back to 3. Do this with multiple examples - +2 and then -2 etc to show that if you add and then subtract the same number you get back to where you started. 3. Use a place value board to create/draw 2-digit numbers – cut up the number track to randomly pull a number from 1-50. Roll a dice marked +1, +2, +3, -1, -

	2, -3 to add or subtract 1,2,3. Mark this onto the drawings. Use one colour to show the start number. Use the second colour to show add or subtract. Model how to write the calculation. Later they may be able to do for themselves .
Games	**Game/ hands on activity** Use a number square from 1-50. You also need a dice marked +1, +2, +3, -1, -2, -3. Players start on 25 and follow the roll of the dice. The winner is the first player to go off the edge of the board. Watch for what happens at the end of the rows of the grid, especially when counting backwards.
Word Problems	**Word Problems** These are all set out as Bett lines – read one line at a time and bet what the question is asking. **Problem Solving Book Level 3** – support with concrete materials – counters and drawing pictures. Track which types of questions the students can and can't do.
Number Problems	**A: Number problems – draw the answer as well as number – pentagon** Model 3+1=4 and 3-1=2, show in the 5 ways • Model - counters • Words – 3 counters plus 1 counter makes 4 counters. • Pictures - draw, use blank number lines to support solving the problems. • Equations - 3+1=4 • Contexts – make a number story. **B: Retrieval and interleaving practice tasks – Level 3 book**
Games	**Game/ hands on activity** Magic Tens game: You will need a place value board and unifix cubes or similar joining cubes e.g. Lego. Take it in turns to spin the spinner (easy 1-3, medium 1-6, hard 4-9) and collect that number of cubes. Once a tower is 10 cubes tall start a new one. The winner is the first player to get 50 (5 completed towers).

Reference cards Module 1

Create a pack of reference cards with the students. Copy the front and then on the back personalize to the student whilst incorporating the learning.

Front	Back
Add	Join 2 groups together. 🐧🐧🐧 and 🐧🐧🐧🐧🐧
Subtract	Remove some items from a group. 🐧🐧🐧🐧🐧🐧
Count on	Start on a number and count on an agreed number of counts more. 6 count on 2 would be 7, 8
Count back	Start on a number and count on an agreed number of counts back. 6 count back 2 would be 5,4
100 square	100 numbers arranged in a square shape.
Missing number	4 plus ? = 10 The missing number here is 6. The missing number can be shown in a number of ways: ? ☐ and later on, letters – as in algebra

Module 2 – Add from largest, Subtract by counting back.
Ideas for teacher lesson (1 hour)

	**Thinking task** Keeping a constant difference – use a number track from 1-50 or number line, can you keep 2 counters the same number of spaces apart. For example, if player A is on 4 and player B is on 6, they are 2 spaces apart – it would take 2 moves to get to the other players space. This would work well with 2 players on a number track/line drawn out on the concrete with chalk, or chalk pen on tiles. Chose a constant difference to explore e.g. different by 2 and record all the numbers which have the same difference e.g. 2 and 4, 3 and 5, 4 and 6 etc. If you write each answer on a different card, you can explore ways to organise them logically and mathematically after a number of goes to look for patterns and then any missing options.
	**Subitizing** Use a set of dominoes – pull out a domino Can the student subitize each half of the domino and explain how they know? Extend by adding the 2 sides of the domino and discuss the different strategies that can be used e.g. count on from the largest number, doubles, near doubles, add a small number, adjust from a known number.
	**Counting – patterns and objects** 1. Count to and from 50 using a number line, ruler, or tape measure – point to the number as you say it. 2. Write in the numbers from 1-50 on a blank 100 square with only the tens numbers completed. Look for patterns horizontally and vertically and then on the diagonally. Teacher to add numbers past 50 if needed to prove or disprove a possible pattern.
	**Number sense focus area explain and model** **Examples and nonexamples** In this module students are learning to count forwards and backwards along the number line to solve addition and subtraction problems. Can they do this using their own internalised number line? This is the long-term goal to be developed. 1. Teach students that 8-3=5 but it can also be 5+3=8 or 8-5=3 or 3+5=8. Use visuals to show this. This understanding of part, part, whole is an important skill. Make many similar examples and explore the different problems you can create. Cuisenaire rods and Numikon are good ways to visually illustrate this. Counters can also be used and rearranged – use 2 different colours– avoid easily confused colours e.g. red/green.

© Copyright 2024 Mathtastic: Tracy Ashbridge. All rights reserved

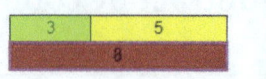 

2. Draw 2 numbers from a pack of playing cards – add the 2 numbers together. Decide which is the largest and count on from there. Students may need support to understand that they do not need to count the first number again and then count on. Once they trust the count, they can just put the biggest number in their head and count on along the visual number line. Students will need to do this on a visual/concrete number line at the start.
3. Finally, students need to learn to count back to subtract small numbers e.g. 8-2. They can put the number in their head, as before, and then step backwards on their number line.

It is easier to count forwards than backwards in maths. Teach students to solve 8-6 by counting on. If they have a good grasp of #1, they will be able to apply this.

Game/ hands on activity
1. Addition dice – roll two 10-sided dice. Add the 2 numbers using counting on from the largest number. Later you can add all 3 dice, decide on the order to add, and give a reason. E.g.

     5+3+1

     7+5+3 or 7+3 (10)+5

An extension of this would be to roll all 3 dice and pick 2 with the aim of reaching a goal – largest answer, smallest answer, nearest to 5 etc.

2. Go fish/ snap/ memory/ stealing game. You are looking to find a player who has the same problem but written in a different way. 3+5=8 could pair with 5+3=8, 8-5=3 or 8-3=5.

Go Fish

The Stealing Game

Snap

Memory Game

	**Word Problems** These are all set out as Bett lines – read one line at a time and bet what the question is asking. Problem Solving Book Level 3 – support with concrete materials – counters and drawing pictures. Track which types of questions the students can and can't do.
	**A: Number problems – draw the answer as well as number – pentagon** Model 3+1=4 and 3-1=2, show in the 5 ways • Model - counters • Words – 3 counters plus 1 counter makes 4 counters. • Pictures - draw, use blank number lines to support solving the problems. • Equations - 3+1=4 • Contexts – make a number story. Teach vertical and horizontal presentation of equations. **B: Retrieval and interleaving practice tasks – Level 3 book**
	**Game/ hands on activity** For this game students need to know odd and even numbers. This can be easily demonstrated by ten frames. Odd numbers have an odd part. Even numbers are nice even shapes. You may need to build these onto the ten frame to get started with the concept. Odd and evens – roll four 10-sided dice (or roll 1 dice 4 times. Group the dice into odd and even numbers. Choose to only add the odd or even dice. The aim of the game is to get the highest number, so some consideration needs to be given to where the most/largest numbers are.

Reference cards Module 2

Create a pack of reference cards with the students. Copy the front and then on the back personalize to the student whilst incorporating the learning.

Front	Back
Difference	The result of subtracting one number from another. How much one number differs from another. Example: The difference between 8 and 3 is 5
Pattern	Patterns include a series or sequence that generally repeats itself. Example: 2 4 2 4 2 4 (repeating pattern) or 2 4 6 8 10 (sequence).
Largest	The biggest group 🐧🐧🐧🐧🐧🐧🐧🐧🐧 🐧🐧🐧🐧
Smallest	The lowest number in a group. 🐧🐧🐧🐧🐧🐧🐧🐧🐧 🐧🐧🐧🐧
Most	Most/ biggest/ greatest 🐧🐧🐧🐧🐧🐧🐧🐧🐧 🐧🐧🐧🐧
Least	Least/ smallest/ lowest 🐧🐧🐧🐧🐧🐧🐧🐧🐧 🐧🐧🐧🐧

© Copyright 2024 Mathtastic: Tracy Ashbridge. All rights reserved

Module 3 – Rainbow Facts
Ideas for teacher lesson (1 hour)

Thinking Problems	**Thinking task** Making tens – you will need a pile of counters and some ten frames. Ask the student to count the pile of counters. How many are there? How can, they be sure? Encourage them to make groups of tens and then recount to check the count is correct. Repeat this with different groups of items to 50. Reinforce making groups of 10 as a strategy.
Subitizing	**Subitizing** Use the cards up to fingers on 4 hands to subitize numbers to 20. Reinforce counting in 5's for full hands if needed. How many more fingers are needed to make a full hand? Tally marks – link this to hands – 1 complete tally mark is 5.
Counting	**Counting – patterns and objects** 1. Use the 100 square to count in 5's to 50 and then 100. Identify the pattern – what changes in the tens and the ones digits? 2. Practice writing any 2-digit number to 50 when given verbally. Random numbers can be generated by turning over 2 playing cards – adapt to be below 50 as needed, e.g. 8 and 3 could be 38. The teacher says thirty-eight and the student writes 38.
Number Sense	**Number sense focus area explain and model** **Examples and nonexamples** Rainbow facts have already been explored in Level 1 and 2. Here we expand this understanding to other tens numbers – friendly numbers. 1. Use visuals – Rekenrek, Cuisenaire, Numikon to build 2-digit numbers and the visually explore how many more are needed to make the next 10. Here 36 is shown and if 4 more are added this makes the next 10 which is 40. 2. Another way to explore this idea visually is to use a partially completed 100 square to colour the tens and ones, then use a different colour to show the complimentary number to make the next group of 10 so

	37+3+40. Show 37, the complimentary number to the next 10 is 3. 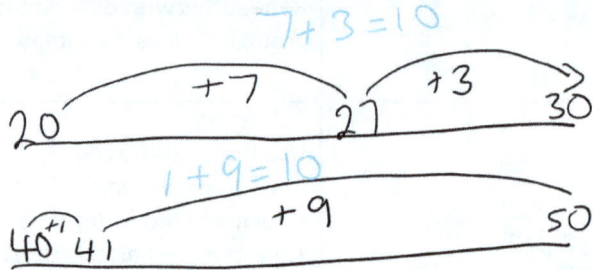 3. Use empty number lines to explore how many more to the next 10. $7+3=10$ $20 \xrightarrow{+7} 27 \xrightarrow{+3} 30$ $1+9=10$ $40 \xrightarrow{+1} 41 \xrightarrow{+9} 50$ 4. Use this to help solve missing number problems e.g. 55+?=60 or 60-?=55
 Games	**Game/ hands on activity** Go Fish, Snap, memory or stealing game. To make a pair you must have 2 cards which add to make a tens number e.g. 22 and 8 or 43 and 7. If you had 43 you would ask to see who has 7 or you could snap/match the pair.
 Word Problems	**Word Problems** These are all set out as Bett lines – read one line at a time and bet what the question is asking. Problem Solving Book Level 3 – support with concrete materials – counters and drawing pictures. Track which types of questions the students can and can't do.
Number Problems	**A: Number problems – draw the answer as well as number – pentagon** Model 3+1=4 and 3-1=2, show in the 5 ways • Model - counters • Words – 3 counters plus 1 counter makes 4 counters. • Pictures – draw, use blank number lines to support solving the problems. • Equations - 3+1=4 • Contexts – make a number story. **B: Retrieval and interleaving practice tasks – Level 3 book**

	**Game/ hands on activity** Follow on game – start with the Go card and match the answer to find the next question.

Reference cards Module 3

Create a pack of reference cards with the students. Copy the front and then on the back personalize to the student whilst incorporating the learning.

Front	Back
Friendly numbers	Numbers which can be used to make it easier to solve problems. 10 is often a friendly number as it is easy to add and subtract.
Tens numbers	Numbers in the ten times table, e.g., 10, 20, 30. These numbers all have a zero in the ones column.
Complimentary numbers	2 numbers which pair to make a tens number 3.g. 37+3=40 37 and 3 are complimentary.

© Copyright 2024 Mathtastic: Tracy Ashbridge. All rights reserved

Module 4 – Add and Subtract 10's.
Ideas for teacher lesson (1 hour)

Thinking Problems	**Thinking task** Count paddle pop sticks, encourage the student to group in tens to make it easier to count at the end. Repeat with other piles of objects to about 50.
Subitizing	**Subitizing** The focus here is on using place value to subitize numbers. Use the cards with bundles of ten (paddle pop sticks) and MAB blocks to quickly represent numbers to 50. Extend the task by asking the students to write the numbers in digits.
Counting	**Counting – patterns and objects** 1. Use a 100 square to support counting in tens to and from 100. Discuss the pattern in the numbers. 2. Write the tens numbers from 10-100 on a blank 100 square, use the patterns to support this. 3. Mark the tens numbers from 1-100 onto a partial number line.
Number Sense	**Number sense focus area explain and model** **Examples and nonexamples** 1. Use MAB, arrow cards, paddle pop sticks and other materials to teach that 30+3=33. 2. Start with numbers from 20-50 as these follow the verbal pattern. Work through a whole tens group in sequence to support understanding e.g. 41, 42, 43,44,45 etc. 3. Include the variations from the same numbers e.g. 30+3=33, 3+30=33, 33-3=30 and 33-30=3. Model this visually with MAB or other place value materials.

© Copyright 2024 Mathtastic: Tracy Ashbridge. All rights reserved

Games	**Game/ hands on activity** • Three bags full place value game – collect 1 sheep to make a bag of wool, and then 10 bags of wool to make a jumper. Further instructions with the game board. • Jump 10 game – you will need a counter marked +1 and jump 10. Each player draws their own number line. Flip the counter and record the calculation onto the number line. The winner is the first to go past 50. You can also play this in reverse, starting at 50 and the first past 0 is the winner.
Word Problems	**Word Problems** These are all set out as Bett lines – read one line at a time and bet what the question is asking. Problem Solving Book Level 3 – support with concrete materials – counters and drawing pictures. Track which types of questions the students can and can't do.
Number Problems	**A: Number problems – draw the answer as well as number – pentagon** Model 3+1=4 and 3-1=2, show in the 5 ways • Model - counters • Words – 3 counters plus 1 counter makes 4 counters. • Pictures – draw, use blank number lines to support solving the problems. • Equations - 3+1=4 • Contexts – make a number story. Teach vertical and horizontal presentation of equations. **B: Retrieval and interleaving practice tasks – Level 3 book**
Games	**Game/ hands on activity** Go fish/snap/memory/stealing game – match up the number and the partitioned number e.g. 33 pairs with 30+3, 45 pairs with 40 +5.

Reference cards Module 4

Create a pack of reference cards with the students. Copy the front and then on the back personalize to the student whilst incorporating the learning.

Front	Back
Tally marks	Groups of marks in 5's to support easier counting. Draw lines for each number and go across to show groups of 5: ||| ￪￪￪￪

Module 5 – Double and Halve.
Ideas for teacher lesson (1 hour)

Thinking Problems	**Thinking task** Use piles of counters or similar small objects and explore what happens when you double or half the group. Record what you notice and wonder. Explore a range of different numbers and look for patterns.
Subitizing	**Subitizing** Ten frames to 50. For this activity use those which are organised pair wise to reinforce doubles. Students should draw on place value from previous sets to support their subitizing.
Counting	**Counting – patterns and objects** 1. Count objects in twos (practice counting in 2's) and explore odd and even numbers. (Count the missing number in your head if needed 2,4,6,7 8, 9 10, 11 12 etc. This can be especially helpful with larger numbers. 2. Count in 2's to and from 50. Use a 100 square to support if needed. Mark the numbers onto the 100 square, what patterns to do you notice? What do you wonder? Teacher to add extra numbers to support wondering if needed.

Number Sense	**Number sense focus area explain and model**
	Examples and nonexamples
	
	1. Reteach the concept of half and double. Model multiple times – depending on the age of the student the link can be explicitly made to multiply and divide by 2. (This is taught in the Australian Curriculum Year 2).
	2. Teach doubling and halving a 2-digit number with concrete materials and visuals (use playing cards to generate numbers and the adjust as needed:
	   a. With no trades – e.g. 22 doubled is 44, 42 halved is 21
	   b. Double numbers with 5 or more in the ones place e.g. 26. Do this with concrete materials: double 6, double 20 and then add together.
	   c. With an odd number in the tens – this will require physical modelling, bundles of 10 are good for this. E.g. 34 halved, halve the 30 and then the 4. If you are using bundles of 10, you can put 10 on each

	half and then unbundle the third ten and split into 5 and 5. The 4 from the ones is easy to halve but you then have to add together all the components from each half: 10+5+2.
Games	**Game/ hands on activity** 1. Playing card doubles – use cards 1-3 only, pull 2 cards and make a number – can you double the number? If it is even, can you halve the number. Include odd numbers when the student is ready for the extra level of challenge. 2. Go fish/snap/ memory doubles – match up the doubles – I have 25, who has 50?
Word Problems	**Word Problems** These are all set out as Bett lines – read one line at a time and bet what the question is asking. Problem Solving Book Level 3 – support with concrete materials – counters and drawing pictures. Track which types of questions the students can and can't do.
Number Problems	**A: Number problems – draw the answer as well as number – pentagon** Model 3+1=4 and 3-1=2, show in the 5 ways - Model - counters - Words – 3 counters plus 1 counter makes 4 counters. - Pictures – draw, use blank number lines to support solving the problems. - Equations - 3+1=4 - Contexts – make a number story. **B: Retrieval and interleaving practice tasks – Level 3 book**
Games	**Game/ hands on activity** Go fish/snap/ memory/stealing halving – match up the halves – I have 50, who has 25?

© Copyright 2024 Mathtastic: Tracy Ashbridge. All rights reserved

Reference cards Module 5

Create a pack of reference cards with the students. Copy the front and then on the back personalize to the student whilst incorporating the learning.

Front	Back
Double	Repeat the same thing twice. 🐧🐧🐧 and 🐧🐧🐧
Half	Make into 2 equal groups. 🐧🐧🐧 🐧🐧🐧
Halve	Make into 2 equal groups. 🐧🐧🐧 🐧🐧🐧
Multiply by 2	Same as double 2 groups of the same number 🐧🐧🐧 and 🐧🐧🐧 2 groups of 3
Times 2	Same as double 2 groups of the same number 🐧🐧🐧 and 🐧🐧🐧 2 groups of 3
Divide by 2	Make into 2 equal groups. 🐧🐧🐧 🐧🐧🐧
X2	Same as double 2 groups of the same number 🐧🐧🐧 and 🐧🐧🐧 2 groups of 3
÷ 2	Make into 2 equal groups. 🐧🐧🐧 🐧🐧🐧

Module 6 – Near Doubles
Ideas for teacher lesson (1 hour)

	**Thinking task** Use the number talks cards to discuss and explore which are doubles, which are not, and which can be solved by near doubles. • 19+19 (double) • 19+20 (near double) • 19+25 (not near double – you could solve by adjusting to 20+24 by moving 1 across) • 20+19 (near double same as 19+20 – which number would you double? – it would be easier to double 20 and adjust down by 1)
	**Subitizing** Rekenrek/abacus place value cards. Recognise the number of tens rows and how many extras – this will reinforce 30+4 = 34 from previous lessons. Use place value to identify the numbers. Write the number in digits. Include place value/spike abacus where each column has a different place value. 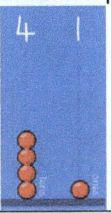
	**Counting – patterns and objects** Count in 2,5,10 to and from 50 with and without a 100 square for reference. Colour code these patterns on a 100 square, what do you notice? Which numbers are in more than one pattern? Why? Look for patterns. Yellow is tens Green is 5's, these are also tens as well Blue – 2's, these also include tens

© Copyright 2024 Mathtastic: Tracy Ashbridge. All rights reserved

1	2	3	4	5	6	7	8	9	10
11	12	13	14	15	16	17	18	19	20
21	22	23	24	25	26	27	28	29	30
31	32	33	34	35	36	37	38	39	40
41	42	43	44	45	46	47	48	49	50
51	52	53	54	55	56	57	58	59	60
61	62	63	64	65	66	67	68	69	70
71	72	73	74	75	76	77	78	79	80
81	82	83	84	85	86	87	88	89	90
91	92	93	94	95	96	97	98	99	100

Number Sense

Number sense focus area explain and model
Examples and nonexamples

1. Teach students how to identify a near double – these are numbers next 2 each other (or 1 apart from each other on a number line). E.g. 23 and 24 are a near double. 24 and 26 could also be considered a near double as you can make 25+25 by moving 1 across – model this with concrete materials to demonstrate this to students.

2. Pull 2 pairs of cards from a pack of cards (1-3 only) and identify near double or not e.g. 31 and 32 yes, 23 and 33 no – allow students to order the cards to make it a near double e.g. 23 could be 32 which would make a near double to 33.
   If it is a near double, solve it, use concrete materials if needed. Lay them in 2 rows side by side to see the double e.g. here 4 and 5 can been seen as double 4 and 1 more.
Use MAB or other place value materials for larger numbers.

3. Once students are able to identify a near double teach them to:
   a) Identify if you can use near doubles?
   b) Which number is easier to double?
   c) Double that number
   d) Adjust to correct to the near double

All the time model this with concrete materials and/or visuals to support the mathematical understanding.

Games	**Game/ hands on activity** Go fish/snap/memory/stealing game – find a near double and solve it
Word Problems	**Word Problems** These are all set out as Bett lines – read one line at a time and bet what the question is asking. Problem Solving Book Level 3 – support with concrete materials – counters and drawing pictures. Track which types of questions the students can and can't do.
Number Problems	**A: Number problems – draw the answer as well as number – pentagon** Model 3+1=4 and 3-1=2, show in the 5 ways • Model - counters • Words – 3 counters plus 1 counter makes 4 counters. • Pictures – draw, use blank number lines to support solving the problems. • Equations - 3+1=4 • Contexts – make a number story. **B: Retrieval and interleaving practice tasks – Level 3 book**
Games	**Game/ hands on activity** Go fish/snap/memory/stealing game find a near double and solve it

Reference cards Module 6

Create a pack of reference cards with the students. Copy the front and then on the back personalize to the student whilst incorporating the learning.

Front	Back
Near double	2 numbers next to each other on the number line e.g., 5 and 6 (or 2 apart e.g., 5 and 7)

Module 7 – Partition by Place Value

Ideas for teacher lesson (1 hour)

Thinking Problems	**Thinking task** How many ways can you break down a number into 2 parts (3 parts)? E.g. 19+ 18+1, 17+2, 3+16 – use counters/ paddle pop sticks. Can the student record as a pattern? If needed, you can write each problem onto a card which can then be reorganised into a pattern.
Subitizing	**Subitizing** Mix together subitizing cards from previous lessons to encourage student to move flexibly between different visual representations. If any particular patterns emerge in their errors, keep to one side for further practice.
Counting	**Counting – patterns and objects** Count to and from 50, then 100, in tens. Use a 100 square to count in tens starting from any number e.g. 3,13,23,33,43 – explore the patterns in the tens and ones numbers. Extend this by drawing counting in tens from any number along a blank number line. (number line showing jumps of +10 from 4 to 14, 24, 34, 44, 54)

Number Sense	**Number sense focus area explain and model**
	Examples and nonexamples
	
	Explore building numbers by partitioning in different ways using tens and ones. Model using tens bundles. e.g. 45 can be:
	    4 tens + 5 ones
	    3 tens + 15 ones
	    2 tens + 25 ones
	    1 ten and 35 ones
	    0 tens and 45 ones.
	Write as a pattern as above.

Games	**Game/ hands on activity** Round about game: In this game students either collect to 50 or subtract from 50 and practice their understanding of place value. Students start on any number, roll the dice, and move. Using place value materials (MAB or paddle pop stick bundles are best), they collect the number they land on. The winner is the first person to break through the target number. The subtraction version is best played with tens bundles so you can unbundle the groups of 10.
Word Problems	**Word Problems** These are all set out as Bett lines – read one line at a time and bet what the question is asking. Problem Solving Book Level 3 – support with concrete materials – counters and drawing pictures. For this module finish any outstanding problems. Track which types of questions the students can and can't do.
Number Problems	**A: Number problems – draw the answer as well as number – pentagon** Model 3+1=4 and 3-1=2, show in the 5 ways • Model - counters • Words – 3 counters plus 1 counter makes 4 counters. • Pictures – draw, use blank number lines to support solving the problems. • Equations - 3+1=4 • Contexts – make a number story. **B: Retrieval and interleaving practice tasks – Level 3 book**
Games	**Game/ hands on activity** Go Fish/ snap/ memory/stealing game – find the same number but partitioned in a different way.

Reference cards Module 7

Create a pack of reference cards with the students. Copy the front and then on the back personalize to the student whilst incorporating the learning.

Front	Back
Partitioning	Partitioning is when you break up numbers using place value e.g., 45 can be partitioned into 4 tens and 5 ones.

Module 8 – Add and Subtract by compensation.
Ideas for teacher lesson (1 hour)

	**Thinking task** Draw 2 cards from a pack- if you add them together, how many different ways could you solve it? E.g. How many different ways can you solve 8+9? • 9 count on 8 • 8 count on 9 • Double 8 plus 1 • Double 9 minus 1 • Change it to 10+7 • Solve 10+8 and adjust by minus 1
	**Subitizing** Mix together subitizing cards from previous lessons to encourage student to move flexibly between different visual representations. If any particular patterns emerge in their errors, keep to one side for further practice.
	**Counting – patterns and objects** Review counting in 1,2,5,10 and from any number in tens to and from 50. Review patterns on the 100 square. | 1	2	3	4	5	6	7	8	9	10
---	---	---	---	---	---	---	---	---	---
11	12	13	14	15	16	17	18	19	20
21	22	23	24	25	26	27	28	29	30
31	32	33	34	35	36	37	38	39	40
41	42	43	44	45	46	47	48	49	50
51	52	53	54	55	56	57	58	59	60
61	62	63	64	65	66	67	68	69	70
71	72	73	74	75	76	77	78	79	80
81	82	83	84	85	86	87	88	89	90
91	92	93	94	95	96	97	98	99	100	 Write any number to 50 in digits given by the teacher – use playing cards to randomly draw a number – adjust as needed.

 Number Sense	**Number sense focus area explain and model** **Examples and nonexamples** Model how to make an equation easier by adjusting the question. Use Cuisenaire/ Rekenrek etc to model this. Example 1: 29 + 14 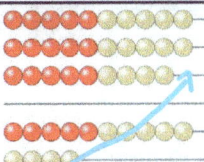 If you move one from the 4 of 14 to the 9 of 29 this simplifies the problem. Example 2: 17+23 can be adjusted so the 3 is paired with the 7 to make a whole 10. 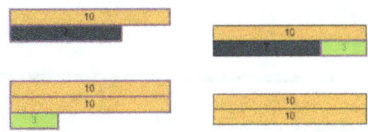 Example 3: 38+6 6 can be broken down into 2 and 4 which can be bridged as 38+2 (makes 40) and then add 4 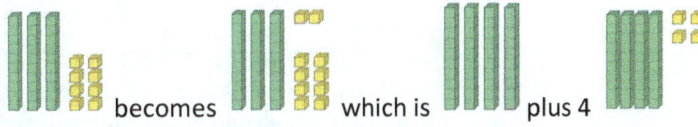 Example 4: You could also show this on a blank number line once students become familiar with breaking down the number to be added on. 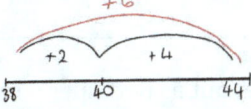 Students need to be flexible in their number sense to understand this so spend time exploring this through concrete and visual materials. Many students will just want to revert to the standard algorithm. The same can also be used for subtraction problems.
 Games	**Game/ hands on activity** Draw 2 cards from a pack of cards to use as the starting number – make this number 10-40. Draw a second single card which will be added to the first number. e.g. 34+8 solve this using bridging e.g. 34+6+2 or another suitable method drawing on number knowledge (not procedures).

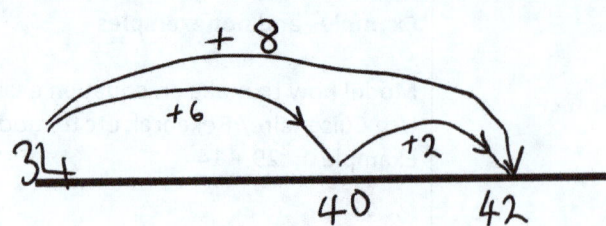

Maybe 34 plus 10 and then minus 2 could work equally as well here.

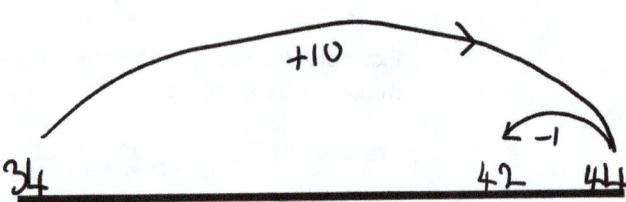

For the second round – keep the single digit number and just change the 2-digit number.
Solve the problem – use the same method as before so students get good at thinking in this way.
e.g.
1. 34 +8
2. 23+8 (how would you adjust this time?)
3. 25+8 (how would you adjust this time?)
4. 12+8 (how would you adjust this time?)

After a few different numbers change the single digit number. Once this is secure for addition, explore with 2 digit subtract 1-digit numbers.

Word Problems
These are all set out as Bett lines – read one line at a time and bet what the question is asking.

Problem Solving Book Level 3 – support with concrete materials – counters and drawing pictures. For this module finish any outstanding problems.

Track which types of questions the students can and can't do.

	**A: Number problems** – draw the answer as well as number – pentagon Model 3+1=4 and 3-1=2, show in the 5 ways • Model - counters • Words – 3 counters plus 1 counter makes 4 counters. • Pictures – draw, use blank number lines to support solving the problems. • Equations - 3+1=4 • Contexts – make a number story. **B: Retrieval and interleaving practice tasks** – Level 3 book
	**Game/ hands on activity** Revisit Three Bags Full and/or Roundabout place value games – record the equations this time and discuss possible strategies for adding or subtracting.

Reference cards Module 8

Create a pack of reference cards with the students. Copy the front and then on the back personalize to the student whilst incorporating the learning.

Front	Back
Adjusting or compensating	Add a little bit too much (or less) to a number and then taking it away (or adding it to the other number. 29+6 becomes 30+5
bridging	Break down the number for addition or subtraction to use a tens number to help.

Level 3

Numbers to 50

Resources

Resources – not included

- Newspaper clipping or similar
- Blank dice marked up - +1, +2, +3, -1, -2, -3
- 2 sided counters
- Unifix or similar joining cubes
- Chalk/ chalk pen
- Dominoes
- Cuisenaire rods
- Pack of playing cards
- Ten-sided dice x 4
- Rekenrek

Lesson Plan

Thinking Problems	**Thinking task**
Subitizing	**Subitizing**
Counting	**Counting – patterns and objects**
Number Sense	**Number sense focus area explain and model** **Examples and nonexamples**
Games	**Game/ hands on activity**

Word Problems	**Word Problems** These are all set out as Bett lines – read one line at a time and bet what the question is asking. Problem Solving Book Level 3 – support with concrete materials – counters and drawing pictures. Track which types of questions the students can and can't do.
Number Problems	**A: Number problems – draw the answer as well as number – pentagon** Model 3+1=4 and 3-1=2, show in the 5 ways • Model - counters • Words – 3 counters plus 1 counter makes 4 counters. • Pictures - draw, use blank number lines to support solving the problems. • Equations - 3+1=4 • Contexts – make a number story. **B: Retrieval and interleaving practice tasks – Level 3 book**
Games	**Game/ hands on activity**

Homework - 15 mins per day

Notes:

Day 1	
Counting	
Number Problems	Number problems booklet Level 3– draw the answer as well as the calculation
Games	
Day 2	
Counting	
Problem Solving	Word Problem solving booklet - Level 3
Games	

Day 3	
Counting	
Number Problems	Number problems booklet Level 3– draw the answer as well as the calculation
Games	
Day 4	
Counting	
Problem Solving	Word Problem solving booklet - Level 3
Games	

Reference cards modules 1-8

Front	Back
Add	Join 2 groups together. 🐧🐧🐧 and 🐧🐧🐧🐧🐧
Subtract	Remove some items from a group. 🐧🐧🐧🐧🐧̶🐧̶
Count on	Start on a number and count on an agreed number of counts more. 6 count on 2 would be 7, 8
Count back	Start on a number and count on an agreed number of counts back. 6 count back 2 would be 5,4
100 square	100 numbers arranged in a square shape.
Missing number	4 plus ? = 10 The missing number here is 6. The missing number can be shown in a number of ways: ? ☐ and later on, letters – as in algebra
Difference	The result of subtracting one number from another. How much one number differs from another. Example: The difference between 8 and 3 is 5
Pattern	Patterns include a series or sequence that generally repeats itself. Example: 2 4 2 4 2 4 (repeating pattern) or 2 4 6 8 10 (sequence).
Largest	The biggest group 🐧🐧🐧🐧🐧🐧🐧🐧🐧 🐧🐧🐧🐧
Smallest	The lowest number in a group. 🐧🐧🐧🐧🐧🐧🐧🐧🐧 🐧🐧🐧🐧
Most	Most/ biggest/ greatest 🐧🐧🐧🐧🐧🐧🐧🐧🐧 🐧🐧🐧🐧

Least	Least/ smallest/ lowest 🐧🐧🐧🐧🐧🐧🐧🐧🐧 🐧🐧🐧
Friendly numbers	Numbers which can be used to make it easier to solve problems. 10 is often a friendly number as it is easy to add and subtract.
Tens numbers	Numbers in the ten times table, e.g., 10, 20, 30. These numbers all have a zero in the ones column.
Complimentary numbers	2 numbers which pair to make a tens number 3.g. 37+3=40 37 and 3 are complimentary.
Tally marks	Groups of marks in 5's to support easier counting. Draw lines for each number and go across to show groups of 5: ‖‖ ‖‖‖‖
Double	Repeat the same thing twice. 🐧🐧🐧 and 🐧🐧🐧
Half	Make into 2 equal groups. 🐧🐧🐧 🐧🐧🐧
Halve	Make into 2 equal groups. 🐧🐧🐧 🐧🐧🐧
Multiply by 2	Same as double 2 groups of the same number 🐧🐧🐧 and 🐧🐧🐧 2 groups of 3
Times 2	Same as double 2 groups of the same number 🐧🐧🐧 and 🐧🐧🐧 2 groups of 3
Divide by 2	Make into 2 equal groups. 🐧🐧🐧 🐧🐧🐧
X2	Same as double 2 groups of the same number 🐧🐧🐧 and 🐧🐧🐧 2 groups of 3

÷ 2	Make into 2 equal groups.
Near double	2 numbers next to each other on the number line e.g., 5 and 6 (or 2 apart e.g., 5 and 7)
Partitioning	Partitioning is when you break up numbers using place value e.g., 45 can be partitioned into 4 tens and 5 ones.
Adjusting or compensating	Add a little bit too much (or less) to a number and then taking it away (or adding it to the other number. 29+6 becomes 30+5
bridging	Break down the number for addition or subtraction to use a tens number to help.

Module 1 Thinking Task - Newspaper
Take a newspaper article and highlight all the numbers – discuss

---

Figures from the Bureau of Meteorology show over 2 metres of rain have fallen in some gauges in the Mossman Gorge region, and 1.9m of rainfall at the Kuranda Railway Station.

The Bairds rain gauge saw 870 millimetres to 9am Monday, the third highest Australian 24-hour rain record, and the heaviest anywhere in Australia since 1958.

Mossman South and Whyanbeel Valley both had over 700mm, which were all time records for the sites.

Meanwhile Cairns has received over 600mm of rainfall over the course of the event.

It's this stamina that has made the event so remarkable, says Professor Turton.

"It's the rates of rainfall, and the fact that it's been over a period of at least two to three days," he said.

How a low-level category cyclone caused Cairns' largest flood in more than a century - ABC News

---

Discussion

How much is 2 metres>

870mm – how much is that? Convert to metres.

24-hour – what is the significance of that length of time

1958 – how do we read that number here? How long ago was that.

[MAB Subitizing cards](#)

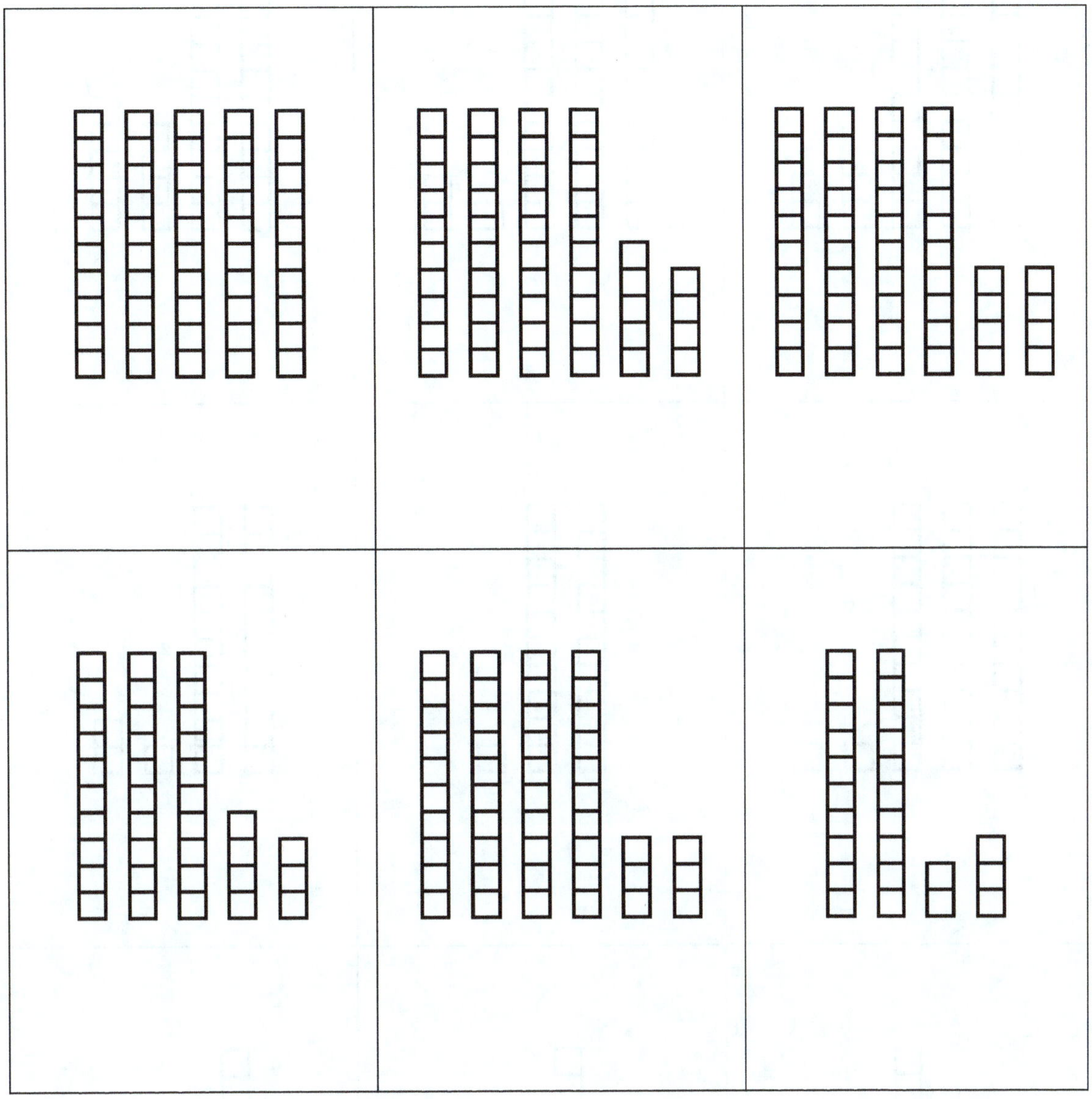

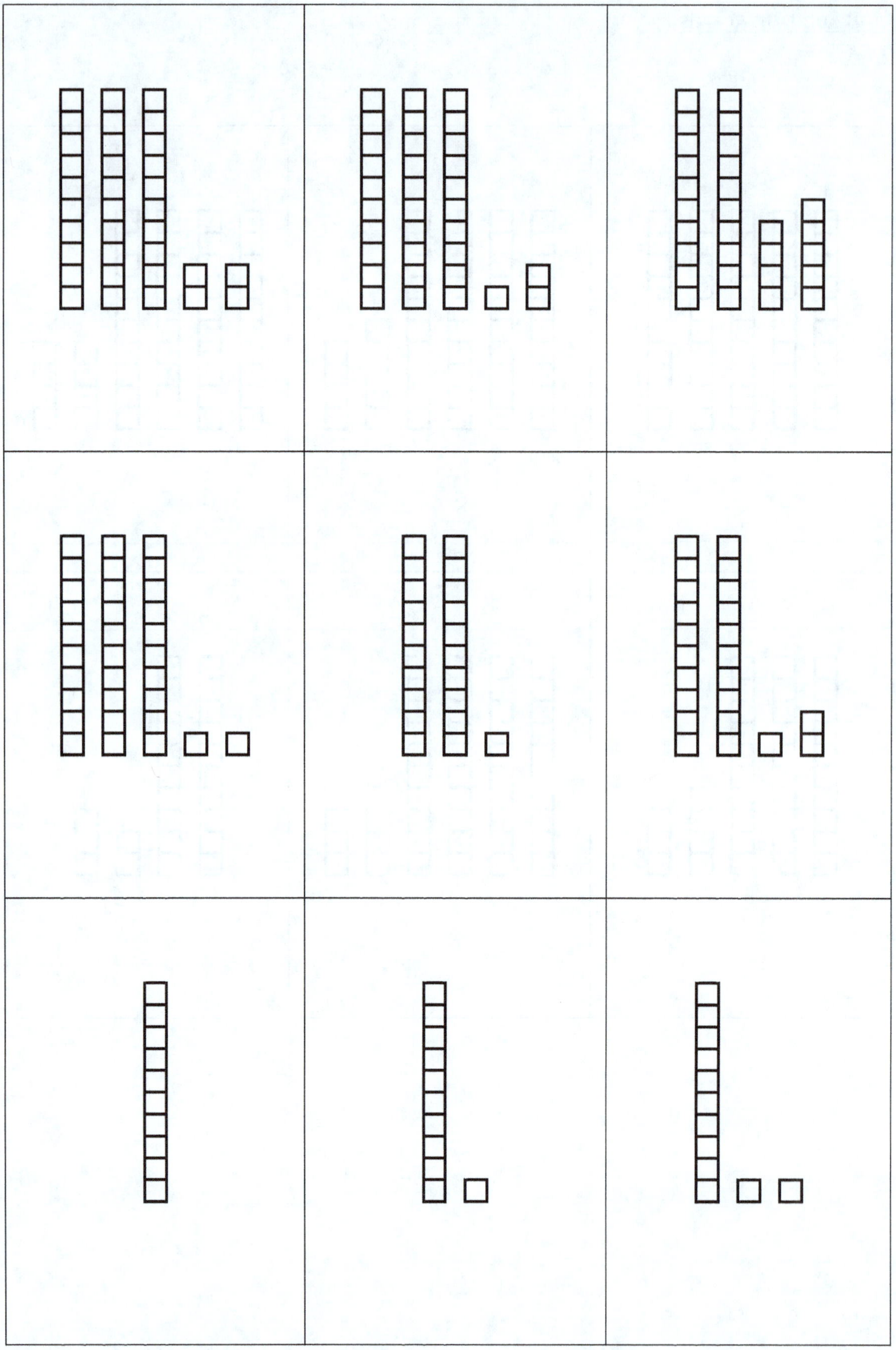

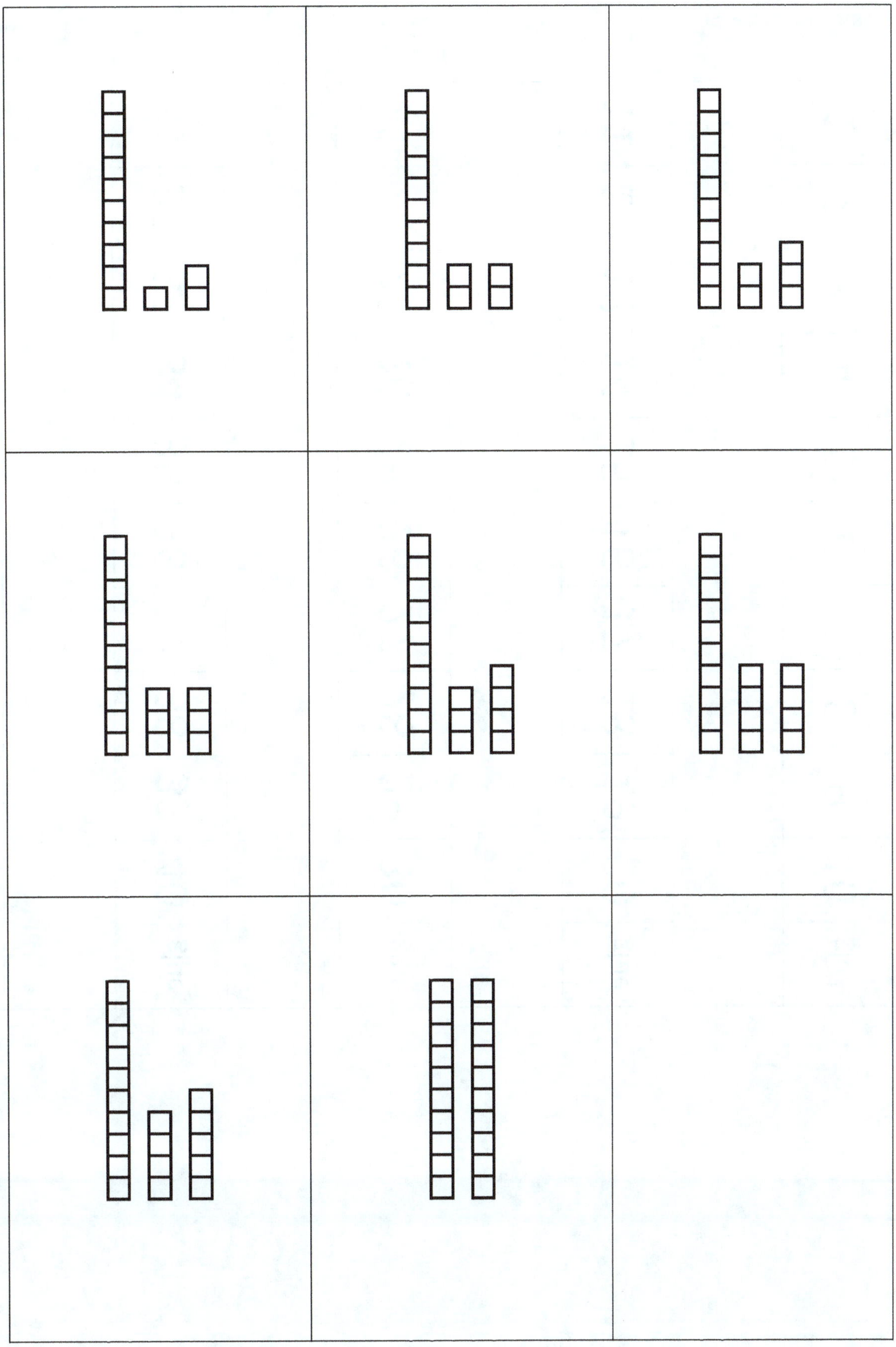

Number Track 1-50

1	11	21	31	41
2	12	22	32	42
3	13	23	33	43
4	14	24	34	44
5	15	25	35	45
6	16	26	36	46
7	17	27	37	47
8	18	28	38	48
9	19	29	39	49
10	20	30	40	50
glue	glue	glue	glue	

Number Tracks 1-50 partially completed.

1		11		21					
		12					33		
							34		
4				24					
5		15		25					
		16		26			37		
		17					38		
9				29					
10		20		30					
glue		glue		glue		glue			

41	42						47	48	49	50	
31						36			39	40	glue
21				25						30	glue
		13	14	15	16					20	glue
1			4	5	6	7					glue

5		15		25		35		45	
10		20		30		40		50	
glue		glue		glue		glue			

Number Track 1-50 blank

Number line 1-50
Overlap the numbers at the end of each row, cut off at 50.

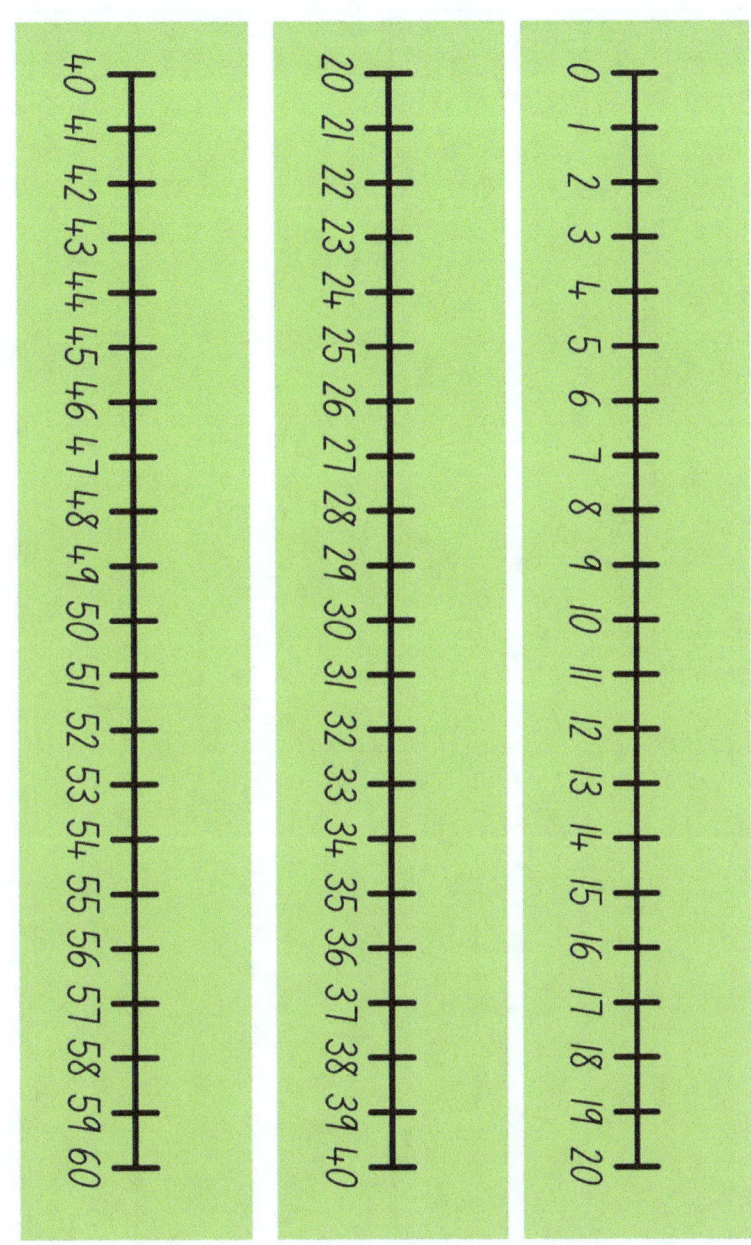

Place Value Board

Tens	Ones

Number Square 1-50 completed.

1	2	3	4	5	6	7	8	9	10
11	12	13	14	15	16	17	18	19	20
21	22	23	24	25	26	27	28	29	30
31	32	33	34	35	36	37	38	39	40
41	42	43	44	45	46	47	48	49	50

© Copyright 2024 Mathtastic: Tracy Ashbridge. All rights reserved

Number Square 1-50 partially completed.

1			4	5	6				10
11			14				18	19	20
21	22	23	24	25	26	27			
31	32			35	36		38	39	40
		43	44	45	46	47			50

Number Square 1-50 blank

Spinners for Magic Tens Game

Easy 1-3

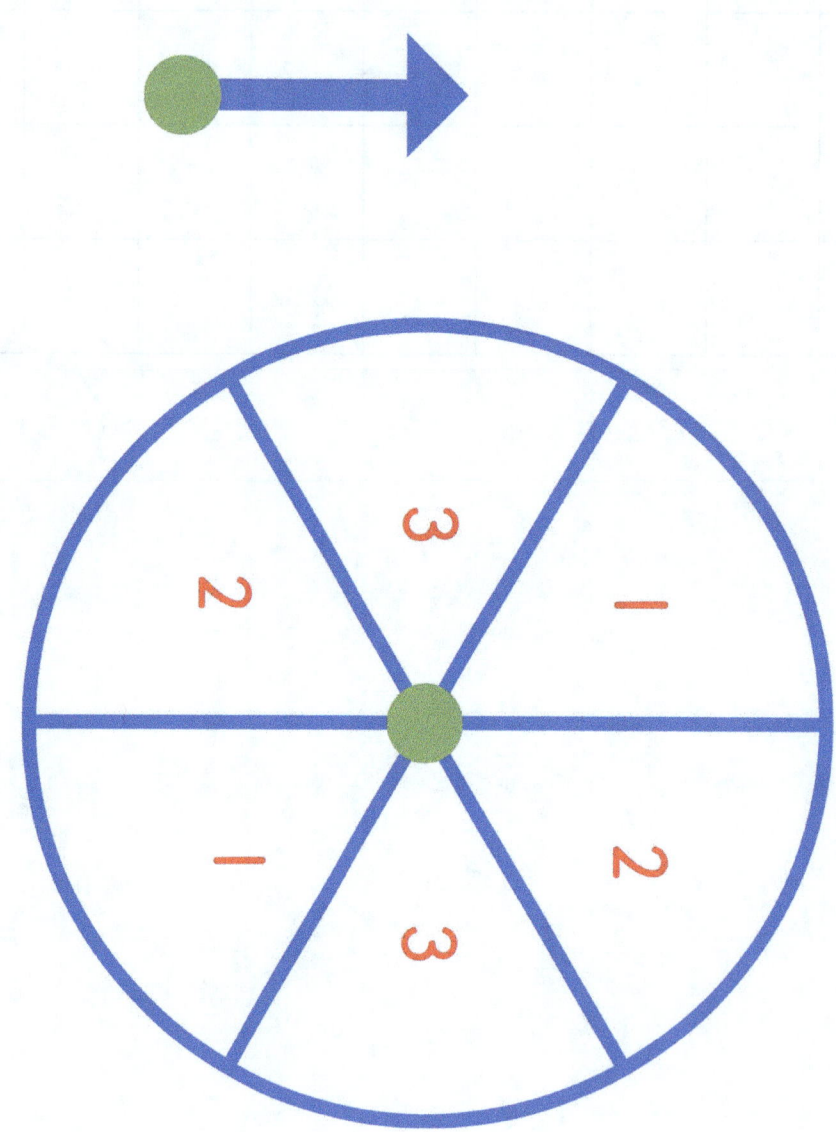

Medium 1-6

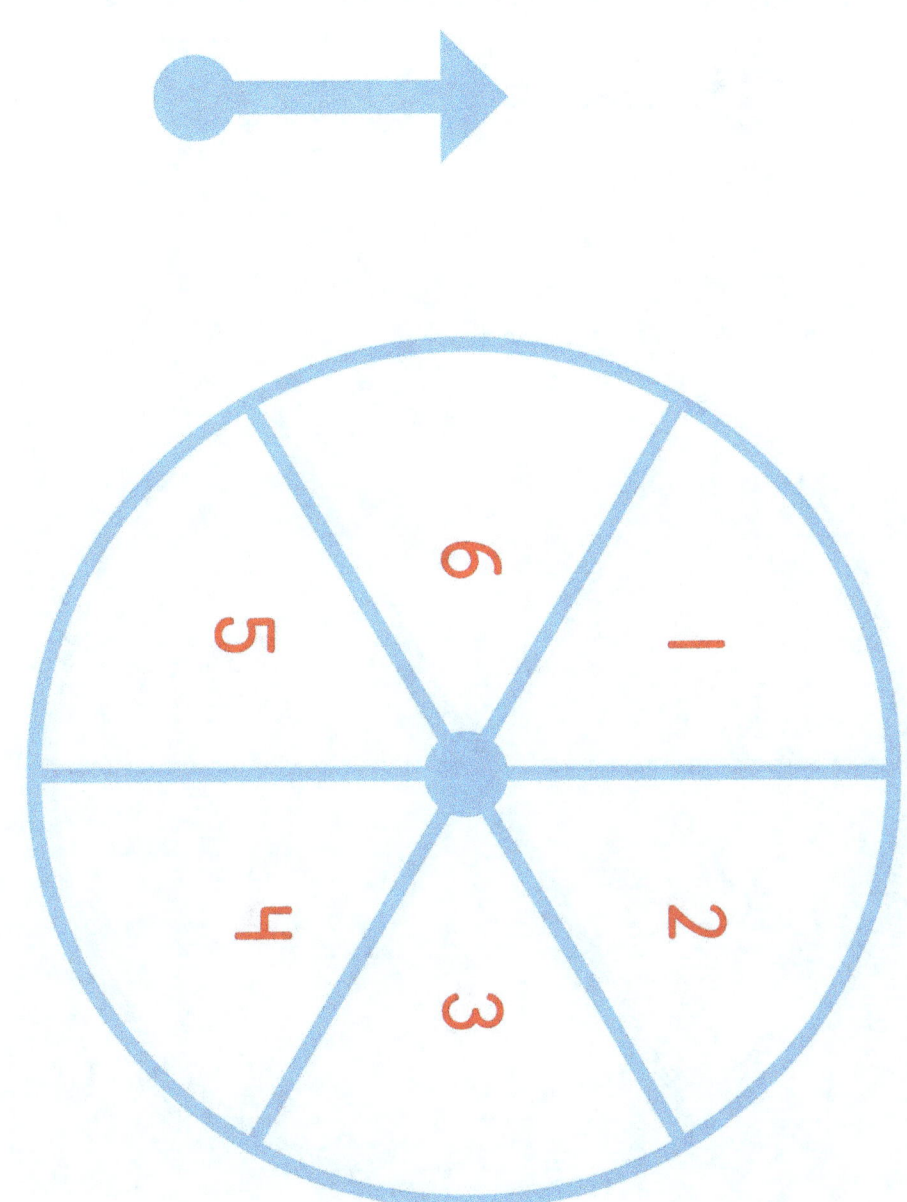

Hard 4-9

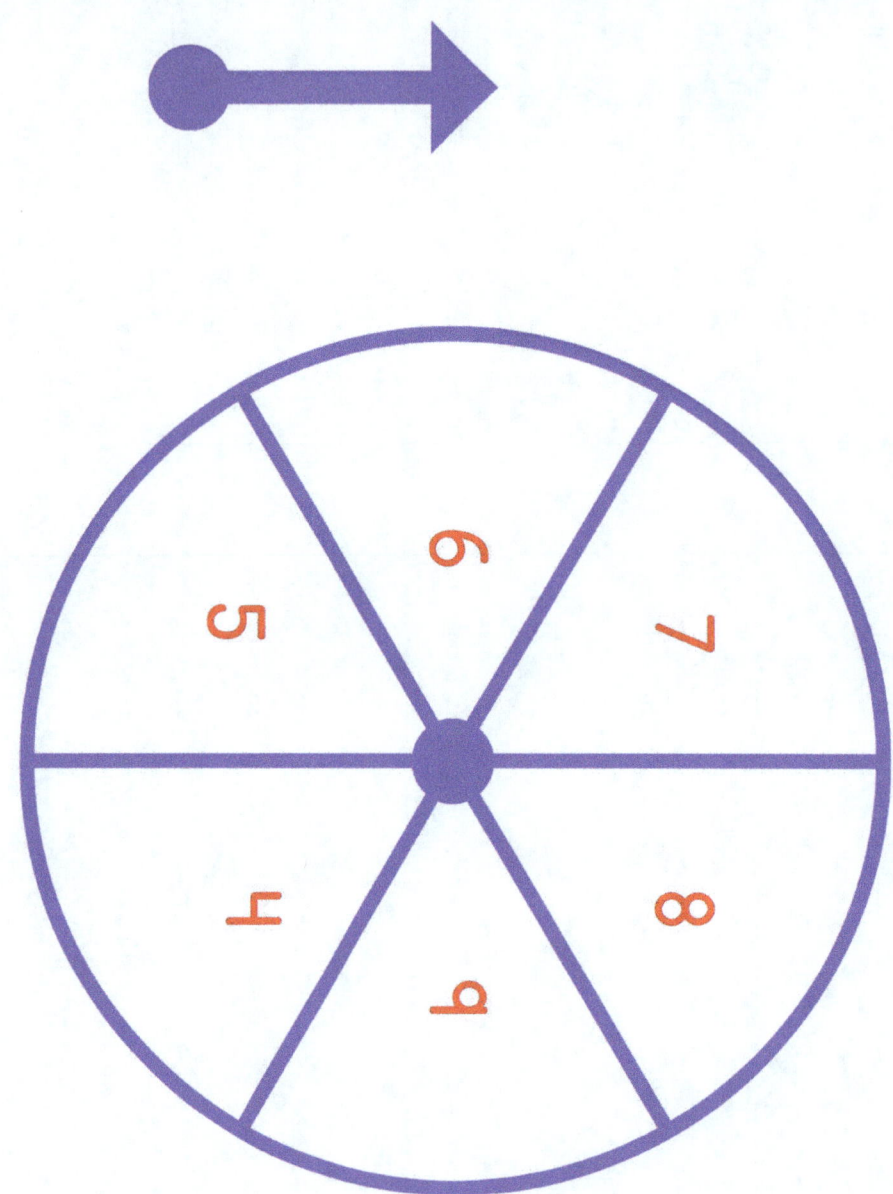

Module 2: Subitizing with dominoes

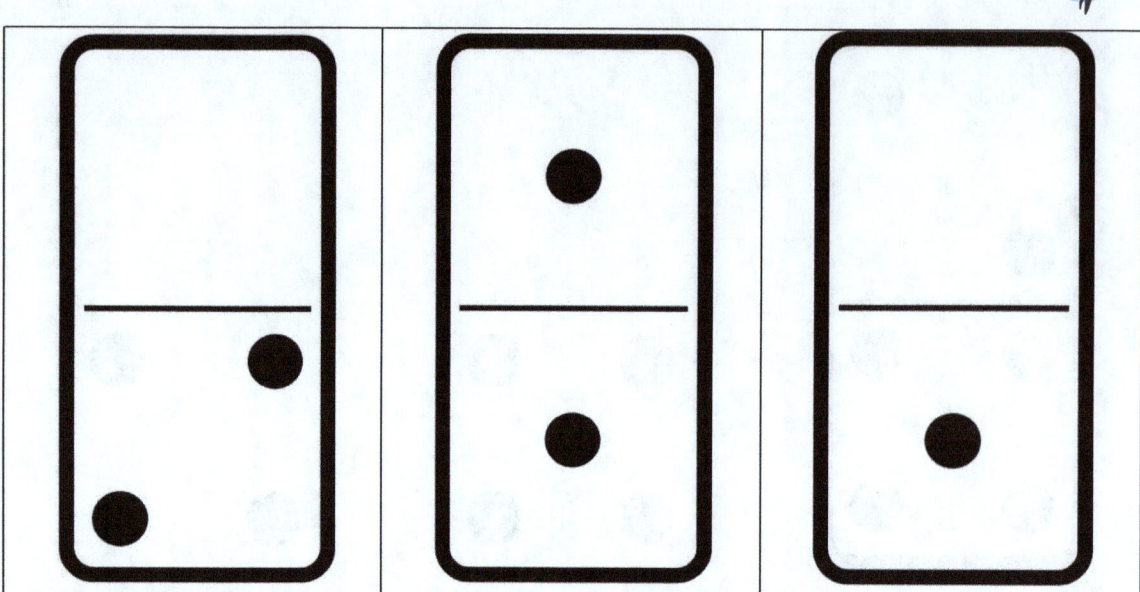

Module 2 – Go fish, snap, memory or stealing game - same question in a different format.

18+3=21	32+5=37	24+2=26
3+18=21	5+32=37	2+24=26

© Copyright 2024 Mathtastic: Tracy Ashbridge. All rights reserved

21-3=18	37-5=32	26-2=24
21-18=3	37-32=5	26-24=2
41+7=48	28+4=32	17+6=23

© Copyright 2024 Mathtastic: Tracy Ashbridge. All rights reserved

7+41=48	4+28=32	6+17=23
48-7=41	32-28=4	23-17=6
48-41=7	32-4=28	23-6=17

18+3=21	32+5=37	24+2=26
3+18=21	5+32=37	2+24=26

Module 3: Blank Ten Frames

Module 3: Finger Subitizing Cards to 20

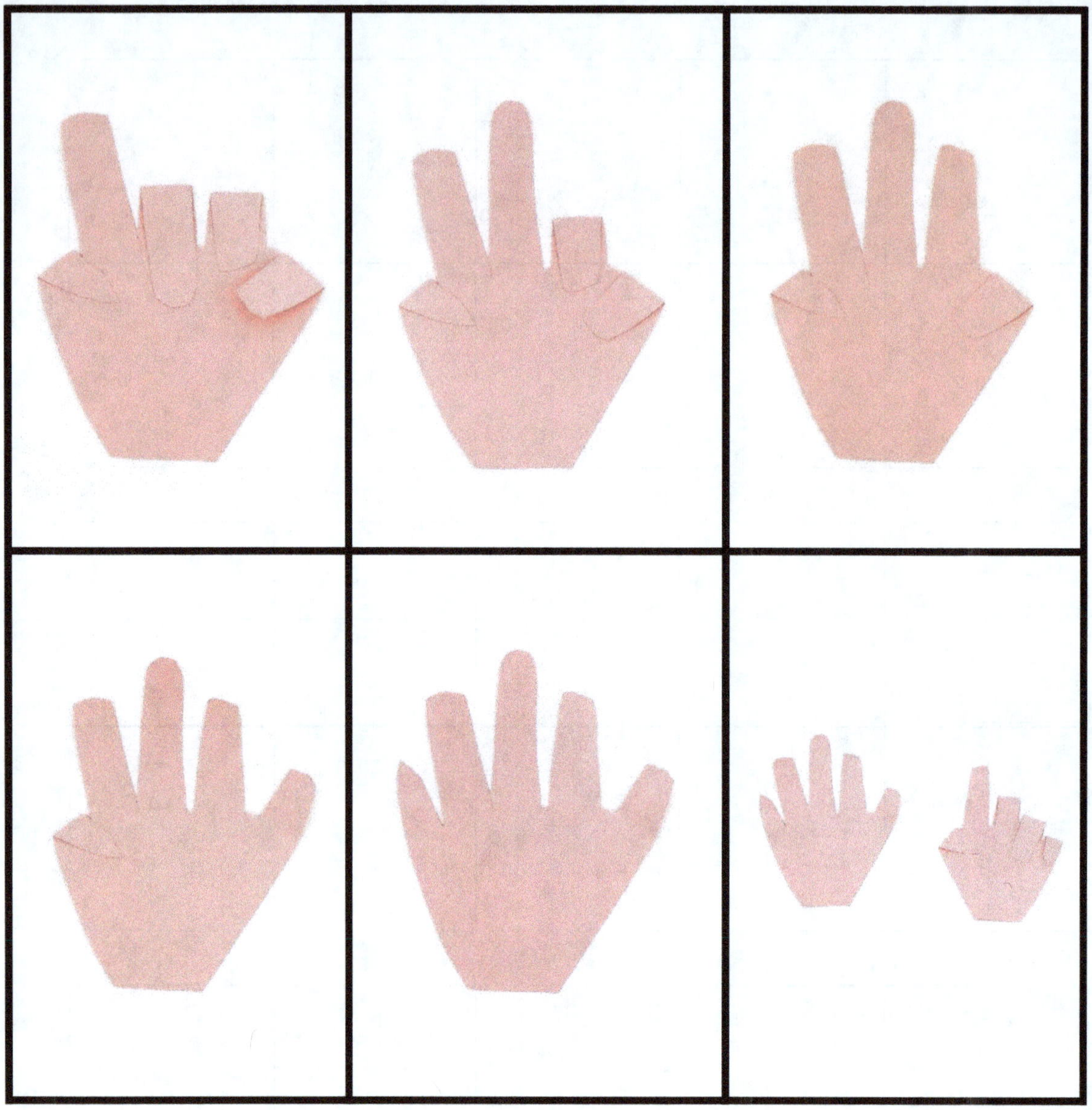

© Copyright 2024 Mathtastic: Tracy Ashbridge. All rights reserved

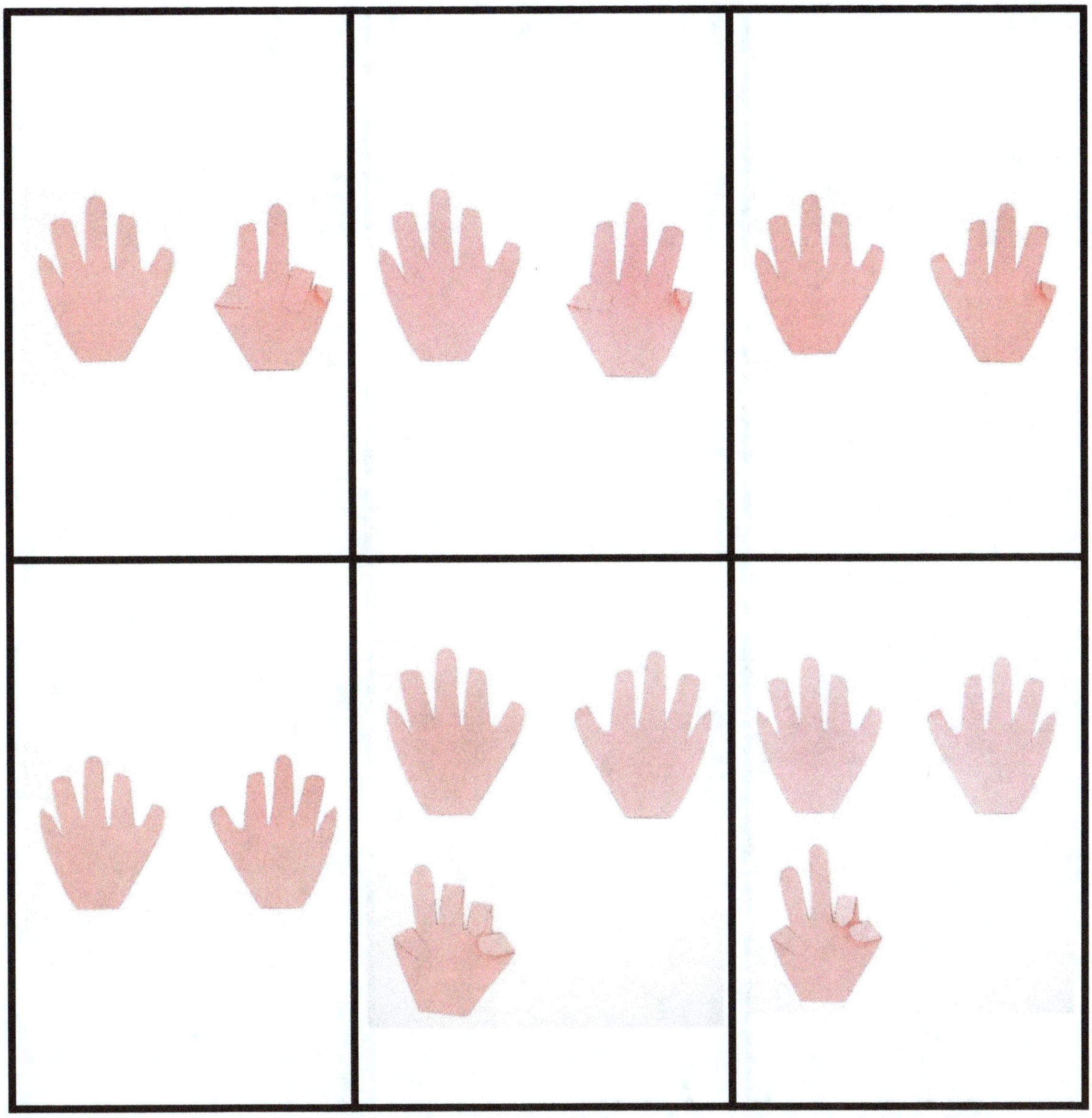

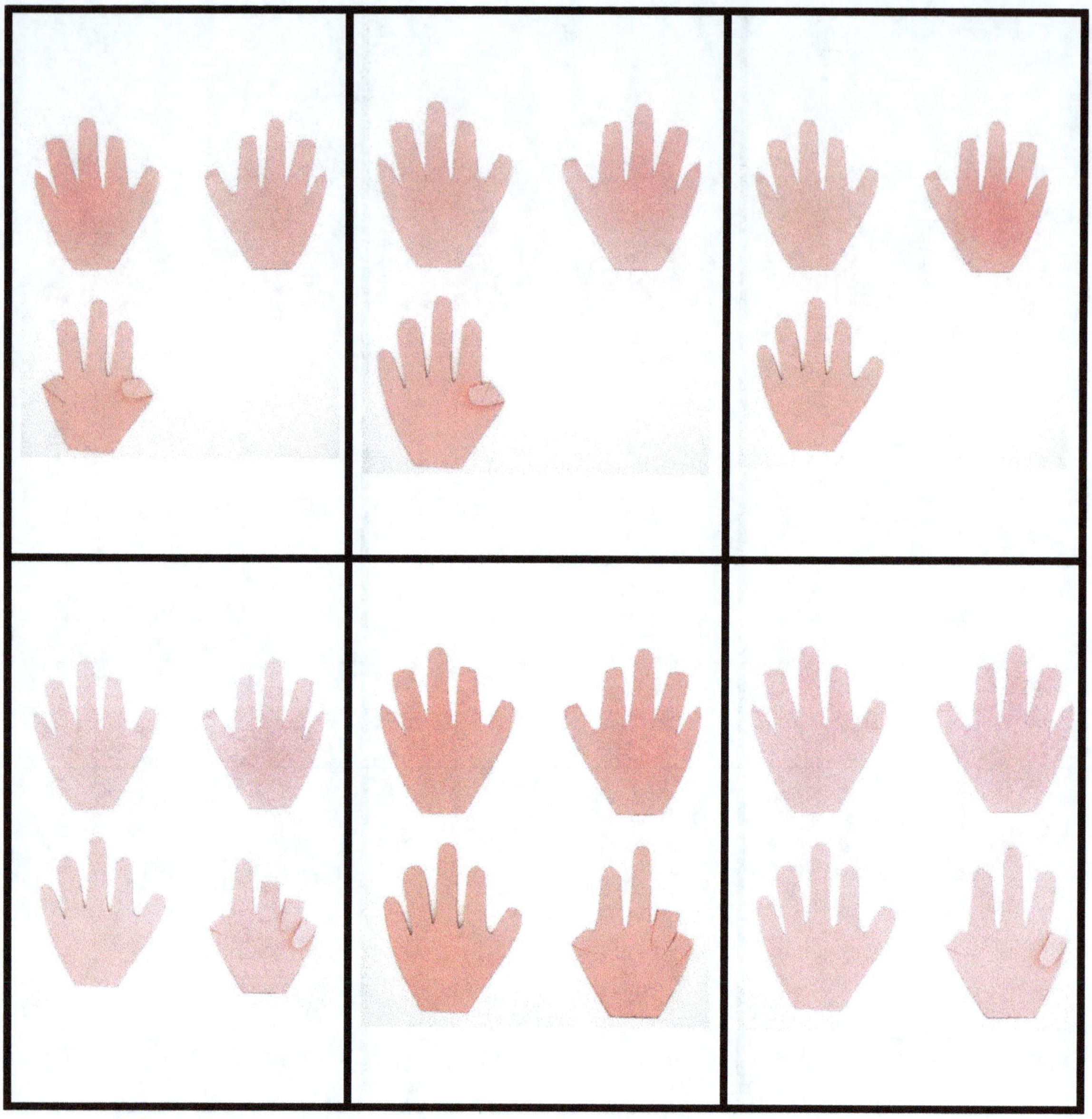

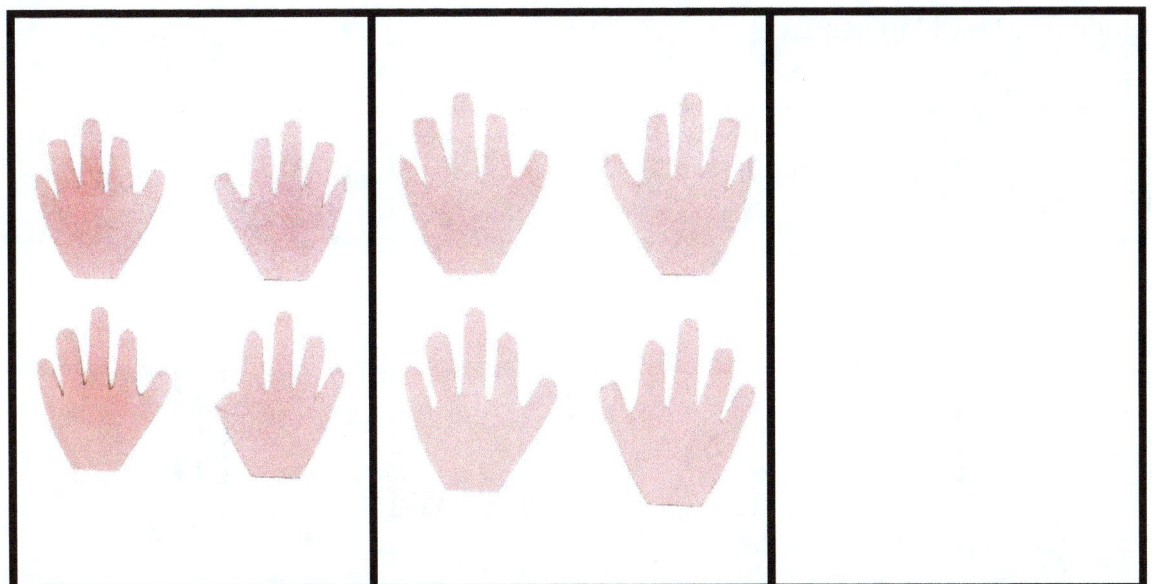

Module 3: Tally Marks to 20

I	II	III
IIII	IIII̸	IIII̸ I

𝍫 𝍪	𝍫 𝍫	𝍫 𝍫𝍫
𝍫 𝍫	𝍫 𝍫 𝍤	𝍫 𝍫 𝍪
𝍫 𝍫 𝍫	𝍫 𝍫 𝍫𝍫	𝍫 𝍫 𝍫

© Copyright 2024 Mathtastic: Tracy Ashbridge. All rights reserved

11	12	13
14	15	

One Hundred Square completed.

1	2	3	4	5	6	7	8	9	10
11	12	13	14	15	16	17	18	19	20
21	22	23	24	25	26	27	28	29	30
31	32	33	34	35	36	37	38	39	40
41	42	43	44	45	46	47	48	49	50
51	52	53	54	55	56	57	58	59	60
61	62	63	64	65	66	67	68	69	70
71	72	73	74	75	76	77	78	79	80
81	82	83	84	85	86	87	88	89	90
91	92	93	94	95	96	97	98	99	100

© Copyright 2024 Mathtastic: Tracy Ashbridge. All rights reserved

One Hundred Square partially completed.

				5					10
				15					20
				25					30
				35					40
				45					50
				55					60
				65					70
				75					80
				85					90
				95					100

© Copyright 2024 Mathtastic: Tracy Ashbridge. All rights reserved

One Hundred Square partially completed.

1			4	5	6	7	8	9	10
11	12	13	14				18	19	20
21				25	26	27			30
31	32	33				37	38	39	
41	42		44	45			48	49	
51					56	57	58		60
61	62	63			66	67		69	70
		73	74	75	76	77	78	79	80
81	82	83			86	87			90
91	92	93				97	98	99	100

One Hundred Square blank

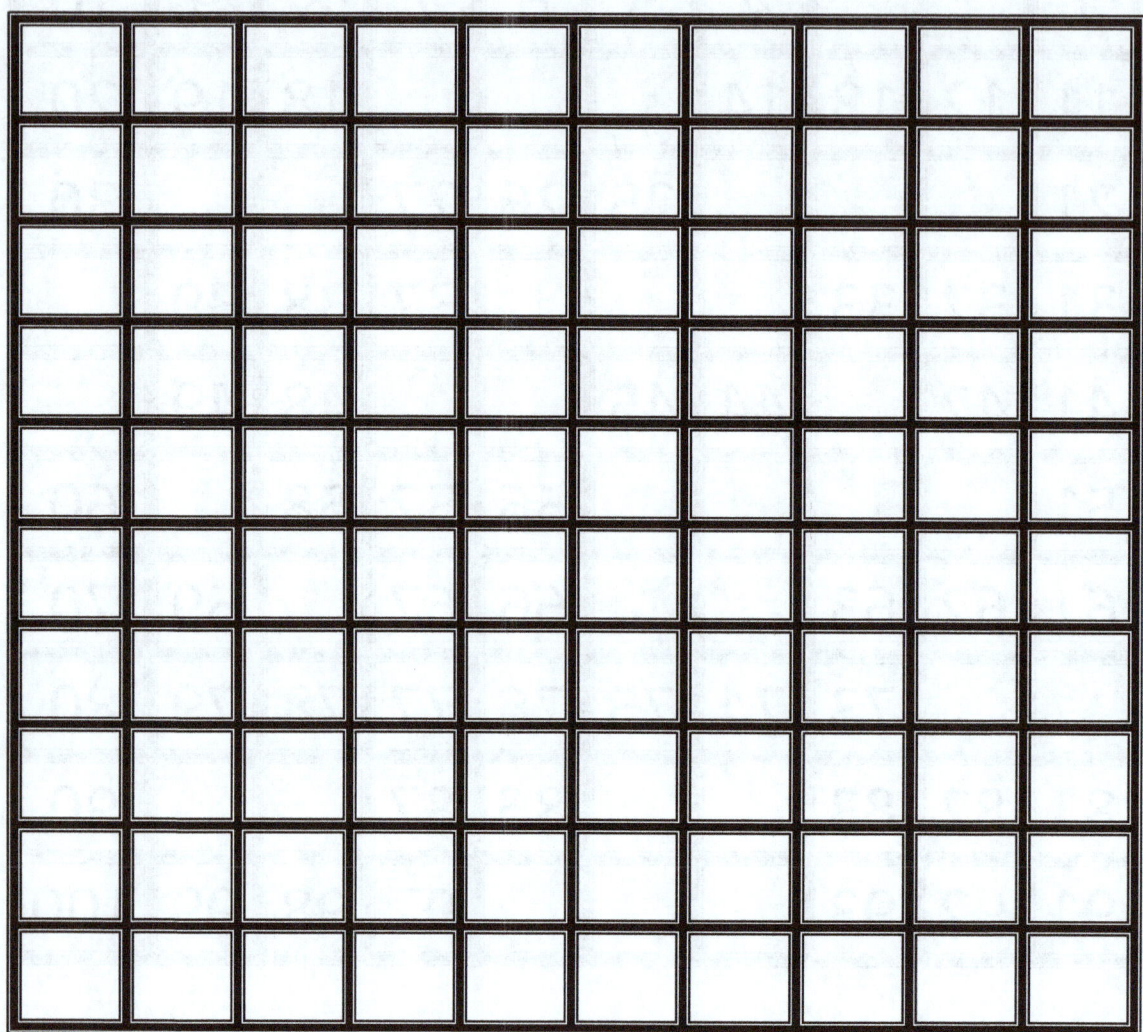

Module 3: Go Fish/ Snap/ Memory/ Stealing game - Rainbow Facts to 50

15	5	45
5	19	1

© Copyright 2024 Mathtastic: Tracy Ashbridge. All rights reserved

49	1	28
2	48	2
27	3	37

© Copyright 2024 Mathtastic: Tracy Ashbridge. All rights reserved

3	46	4
36	4	34
6	14	6

33	7	13
7	22	8
42	8	39

© Copyright 2024 Mathtastic: Tracy Ashbridge. All rights reserved

	1	19	1

Module 3: Follow on game.

Cut across the page to make strips.

Place the cards in line to answer each card.

Start	I have 26+4. Who has 30?
I have 30.	Who has 17+3?
I have 20.	Who has 44+6?
I have 50.	Who has 6+4?

© Copyright 2024 Mathtastic: Tracy Ashbridge. All rights reserved

I have 10.	Who has 31+9?
I have 40.	Finish

Number line 1-100

Overlap the strips to join together.

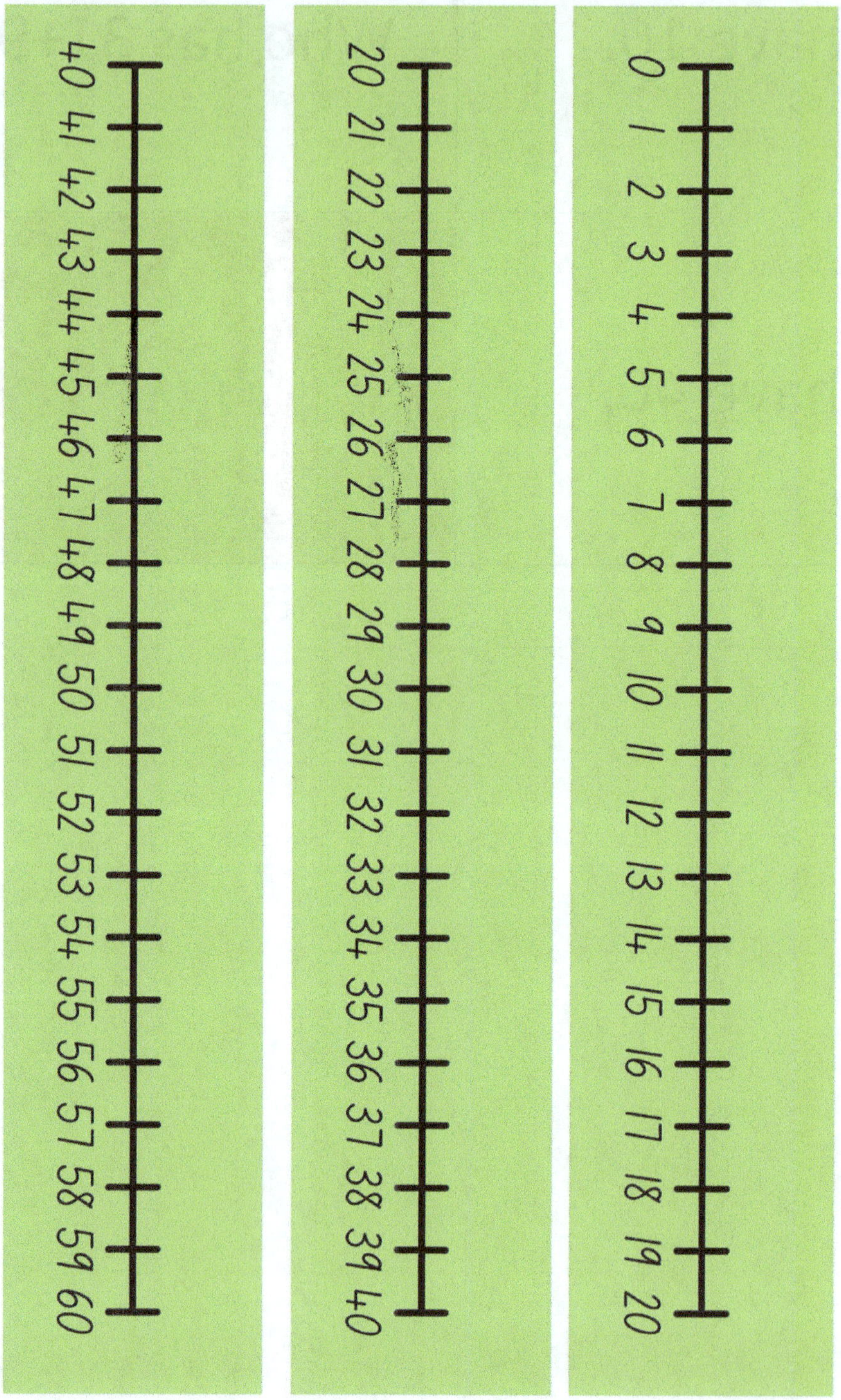

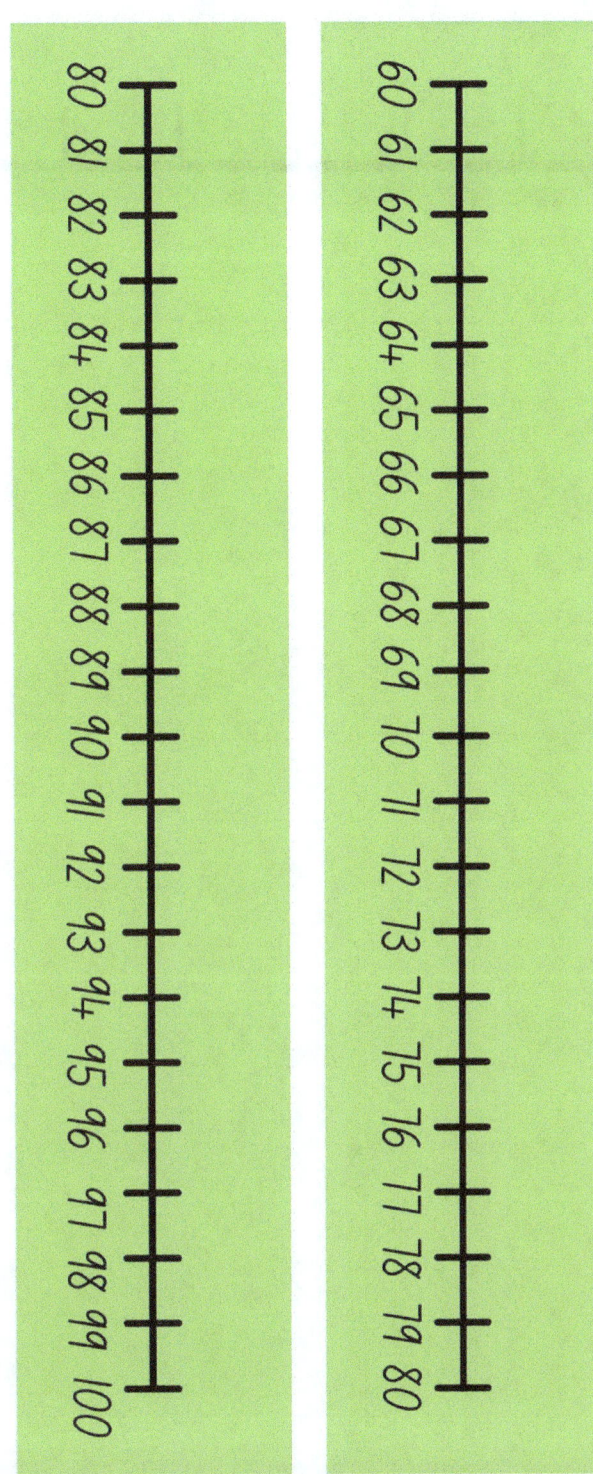

Blank number line

Mark on 0-100 in tens

Module 4: Subitizing paddle pop sticks

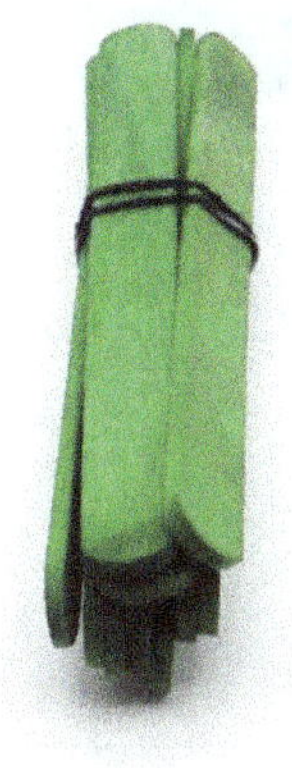

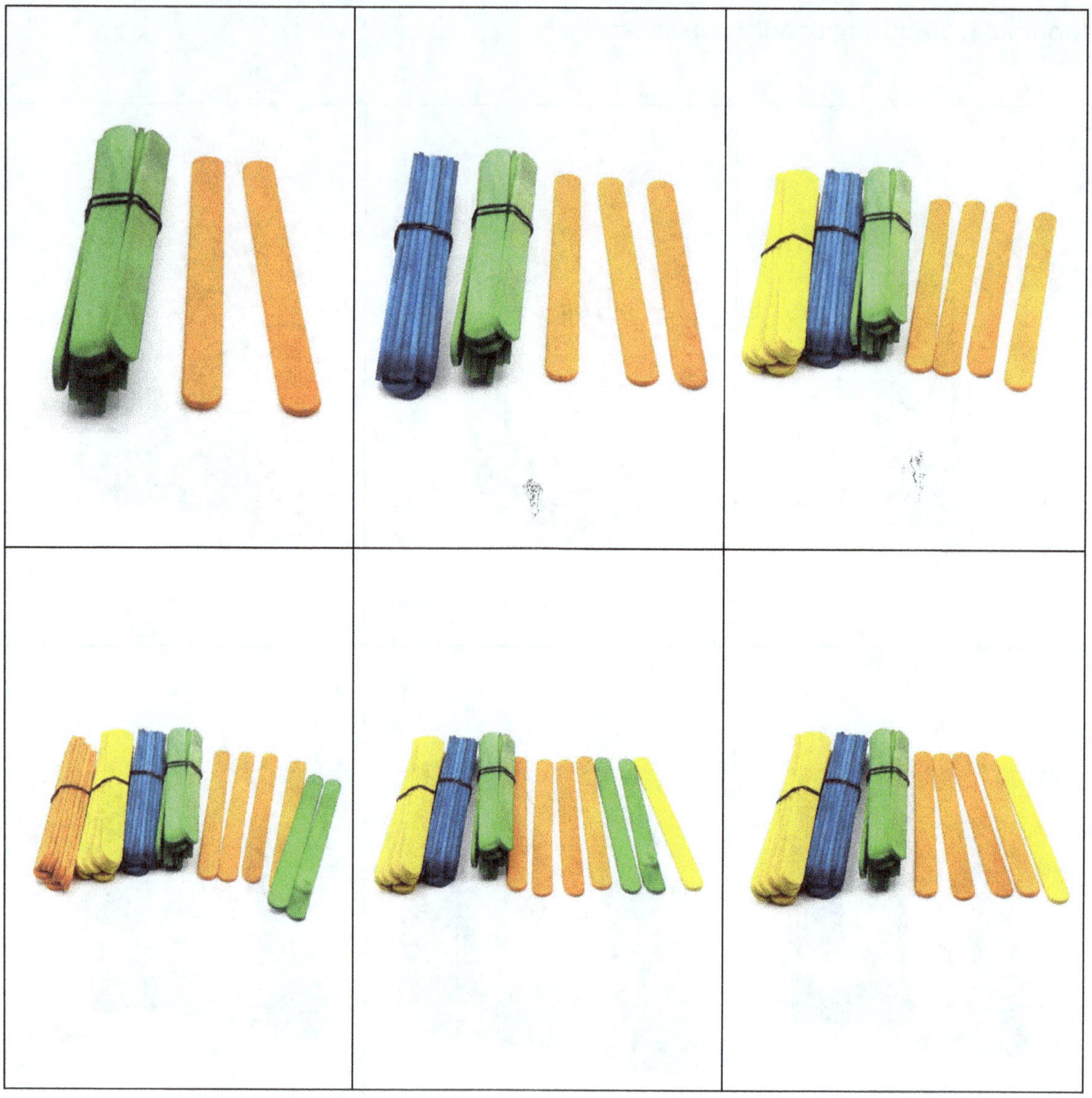

Module 4 - Arrow cards
Cut out the cards and make the arrows pointed as shown.

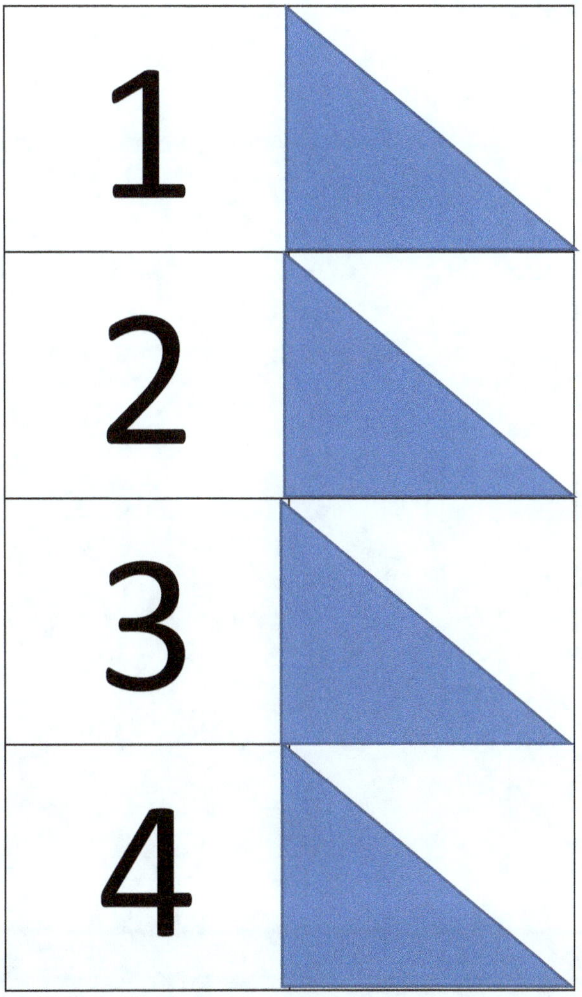

© Copyright 2024 Mathtastic: Tracy Ashbridge. All rights reserved

1	0	
2	0	
3	0	
4	0	
5	0	
6	0	

7	0	
8	0	
9	0	

Module 4: Place Value Game – Three Bags Full

How to play

10 sheep make 1 bag of wool.

10 bags of wool make a jumper.

You will need counters and a six-sided dice.

Decide who will start the game.

Roll the dice and cover that number of sheep.

When all sheep have been covered, 10 sheep = 1 bag of wool

The winner is the first player to collect enough sheep and wool to make a jumper.

Three Bags Full - Place Value Game

Module 4 - Go Fish/ Snap/ Memory/ stealing game – place value.

10+3	13	20+5
25	30+9	39

40+1	41	20+2
22	40+4	44
30+5	35	40+6

© Copyright 2024 Mathtastic: Tracy Ashbridge. All rights reserved

46	10+7	17
40+8	48	50+0
50	10+3	13

© Copyright 2024 Mathtastic: Tracy Ashbridge. All rights reserved

Module 5: Ten frames to 50 – subitizing cards

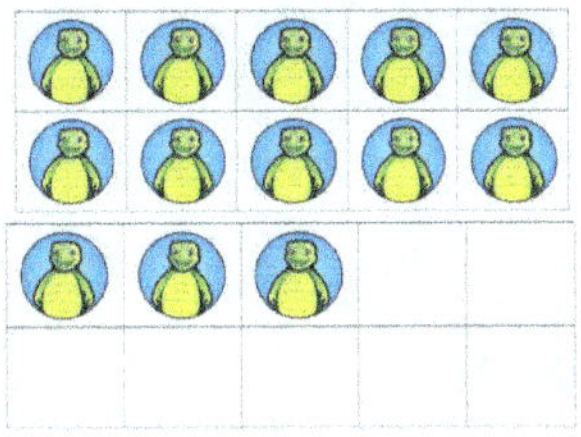

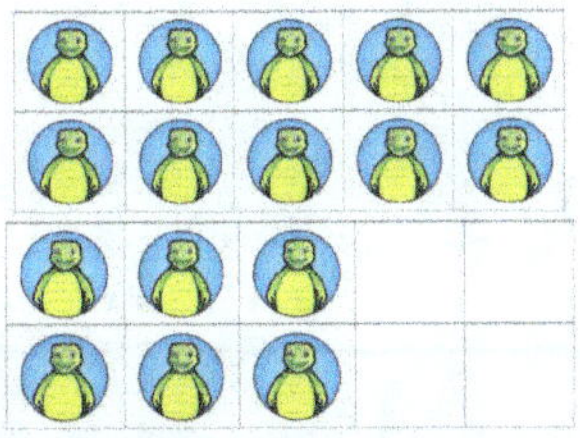

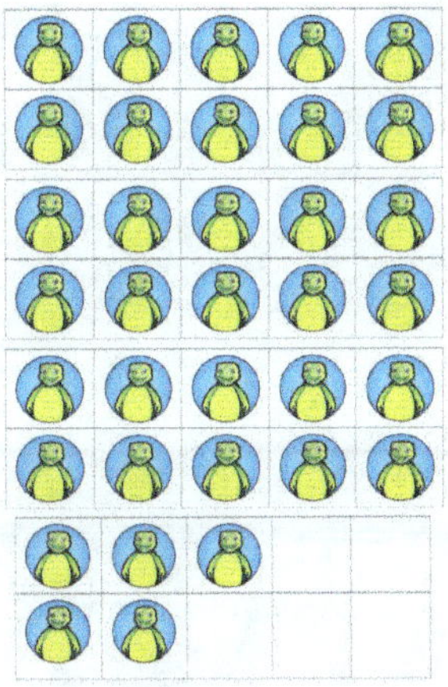

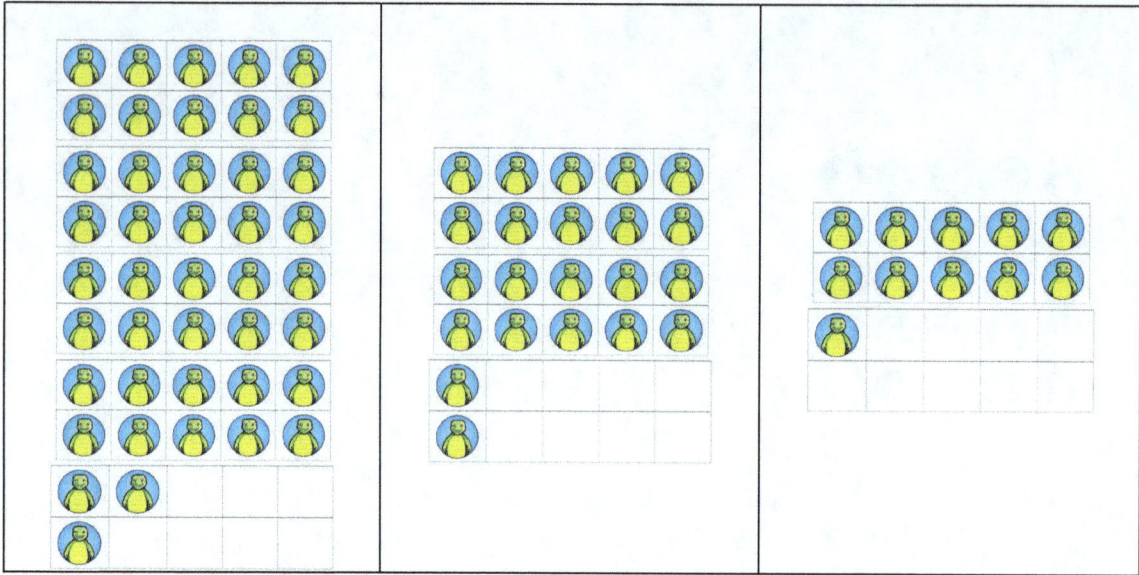

Module 5: Go fish, snap, memory, stealing game doubles.

50	25	48
24	46	23

44	22	42
21	40	20
38	19	36

18	34	17
32	16	30
15	28	14

© Copyright 2024 Mathtastic: Tracy Ashbridge. All rights reserved

26	13	24
12	22	11
20	10	18

9	16	8
14	7	12
6	10	5

8	4	6
3	4	2

Module 6: Number talks cards

Use the number talks cards to discuss and explore which are doubles, which are not, and which can be solved by near doubles.

- 19+19 (double)
- 19+20 (near double)
- 19+25 (not near double – you could solve by adjusting to 20+24 by moving 1 across)

20+19 (near double same as 19+20 – which number would you double? – it would be easier to double 20 and adjust down by 1)

19+19	19+20	19+25
16+17	18+16	15+6

© Copyright 2024 Mathtastic: Tracy Ashbridge. All rights reserved

8+9	8+10	8+13
26+24	24+25	12+25

Module 6: Rekenrek place value subitizing photos

Use how many on the left-hand side.

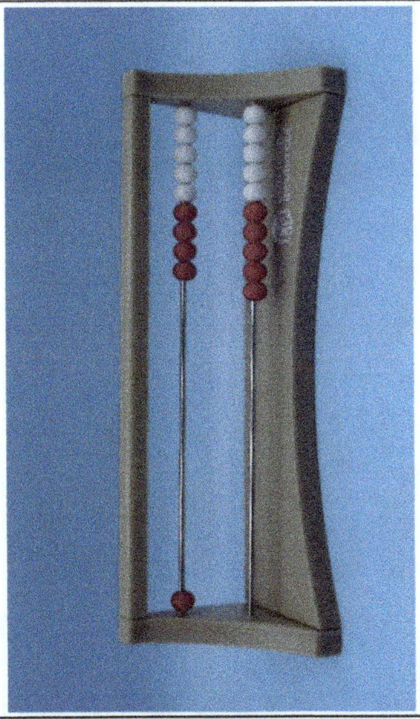

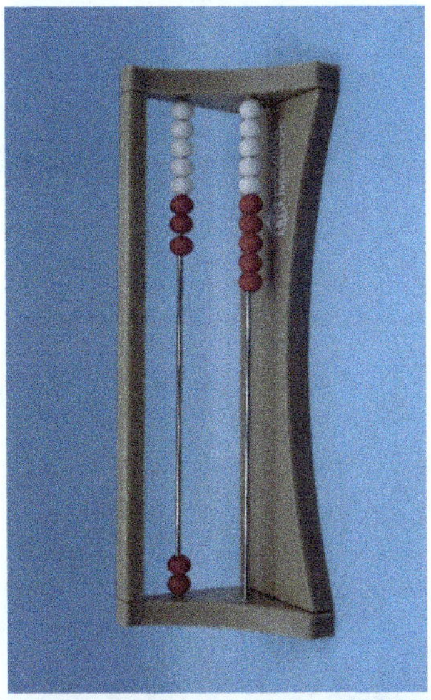

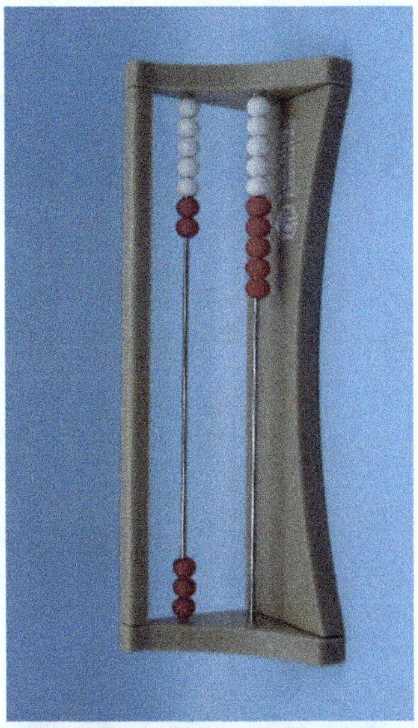

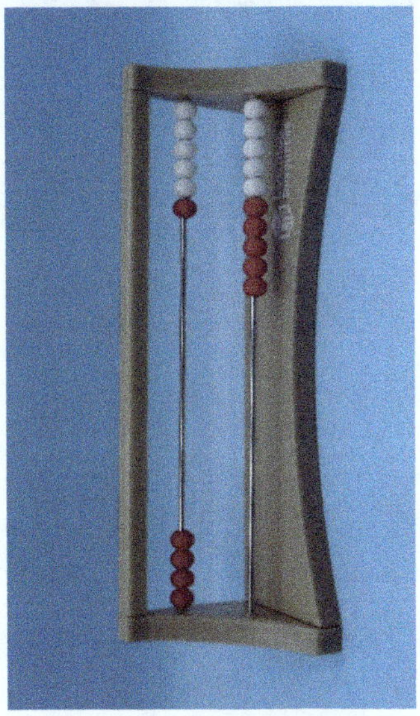

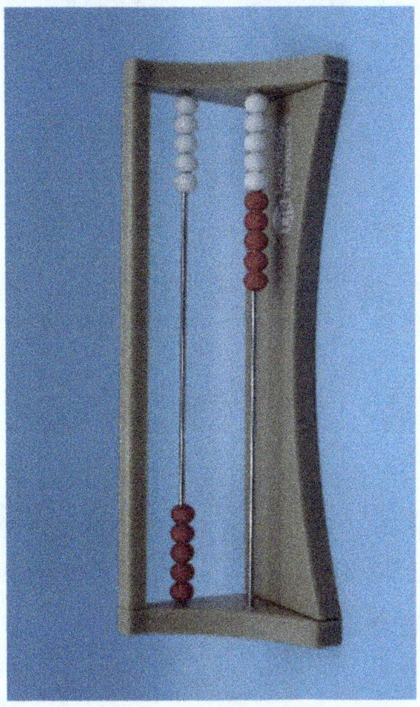

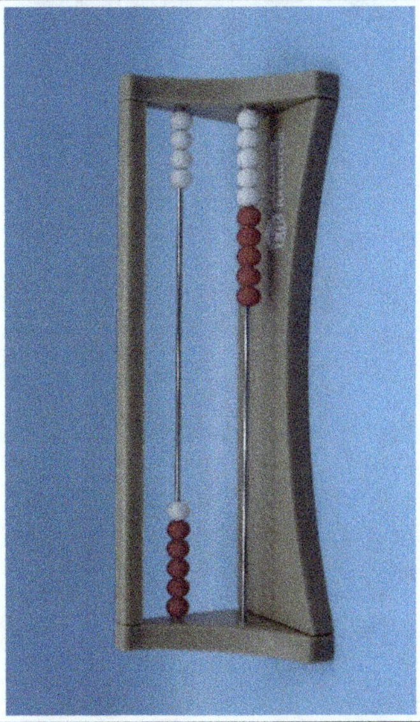

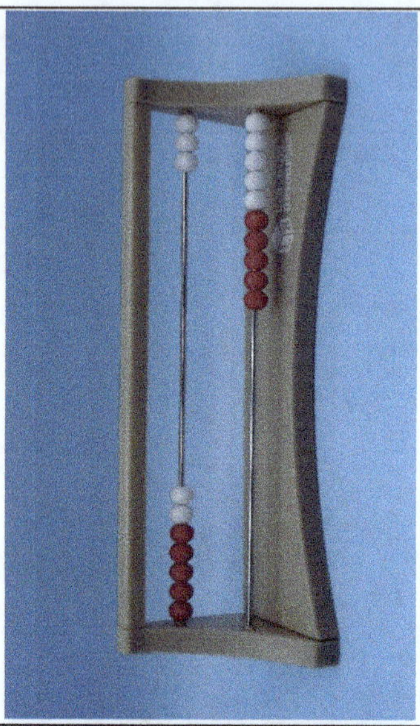

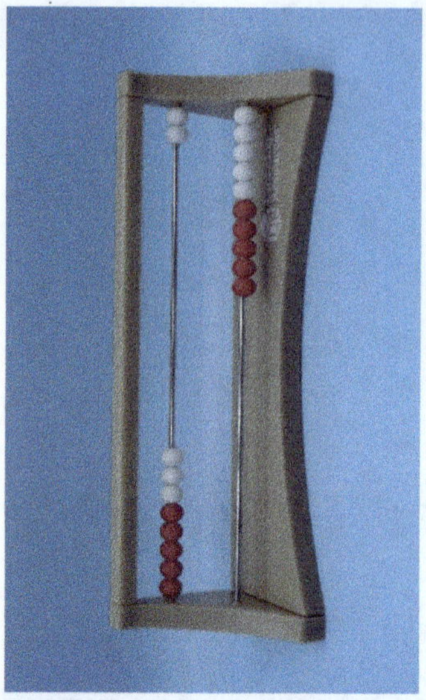

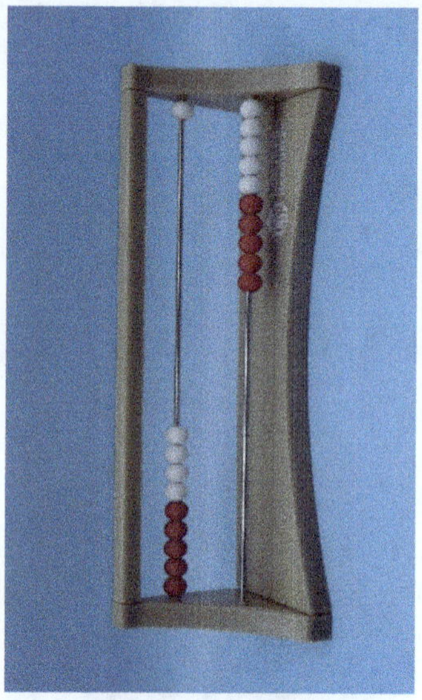

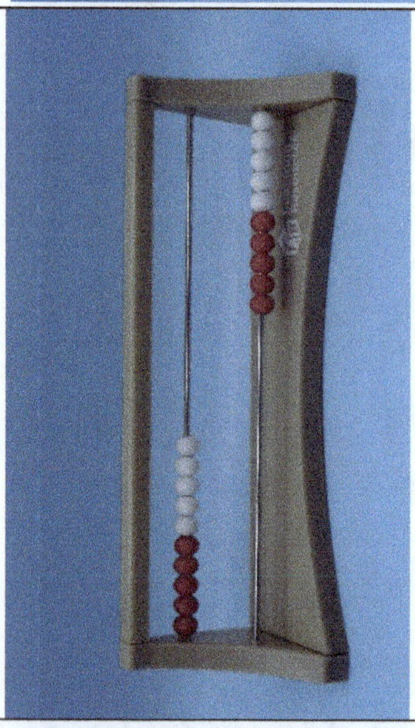

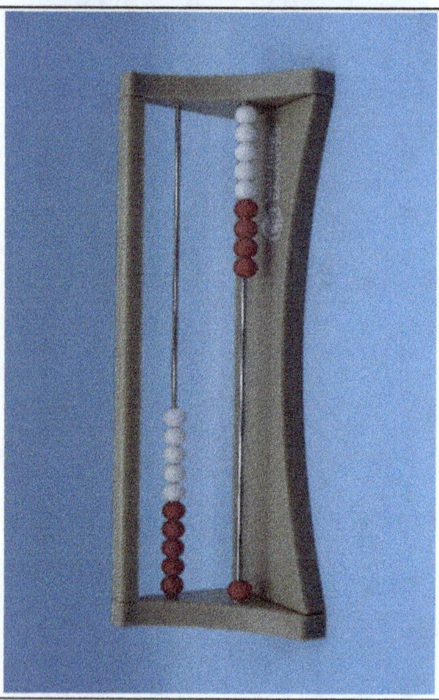

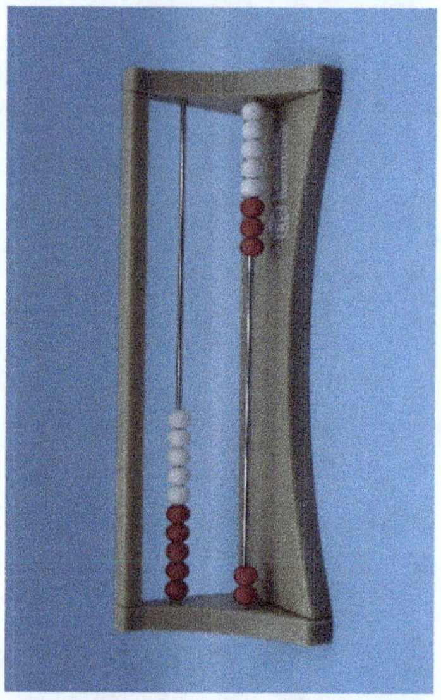

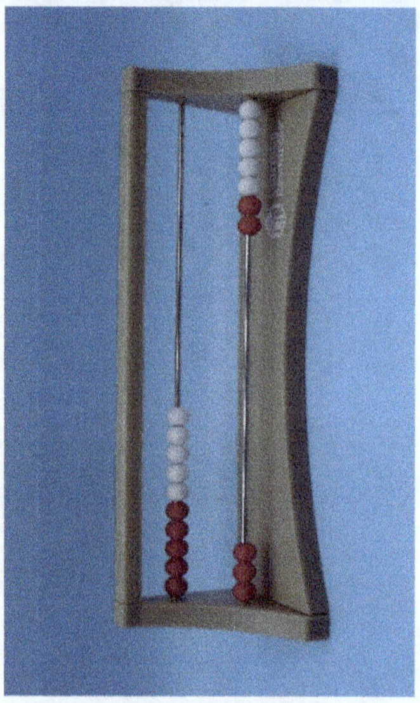

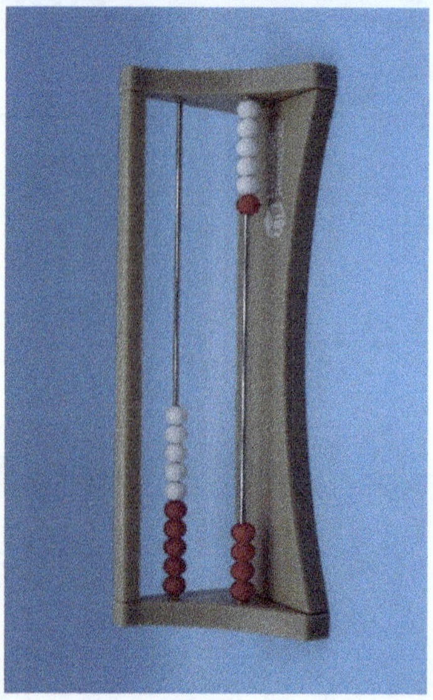

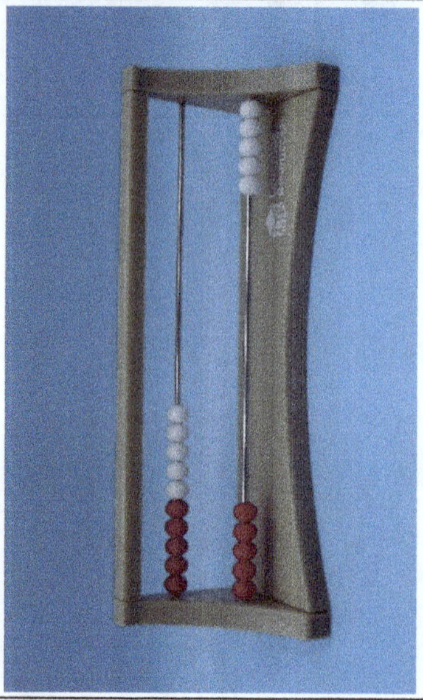

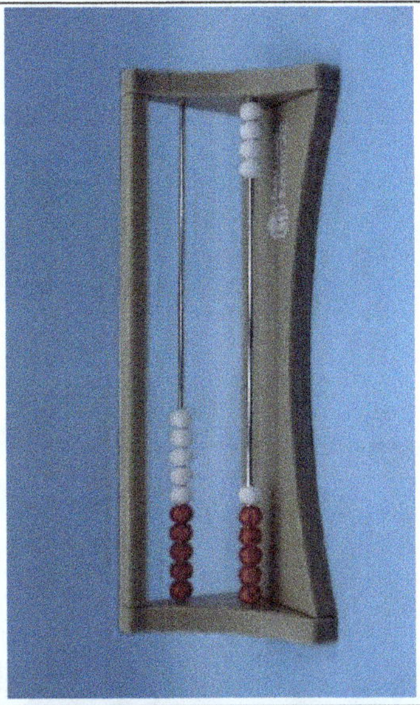

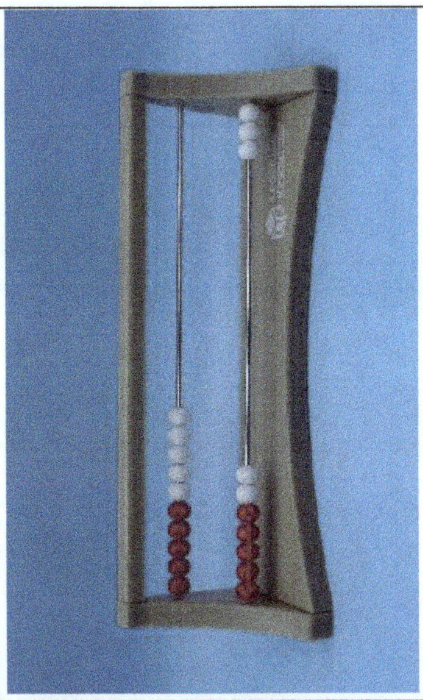

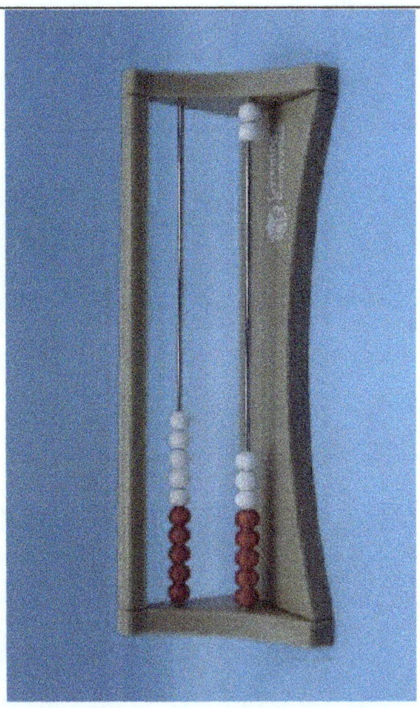

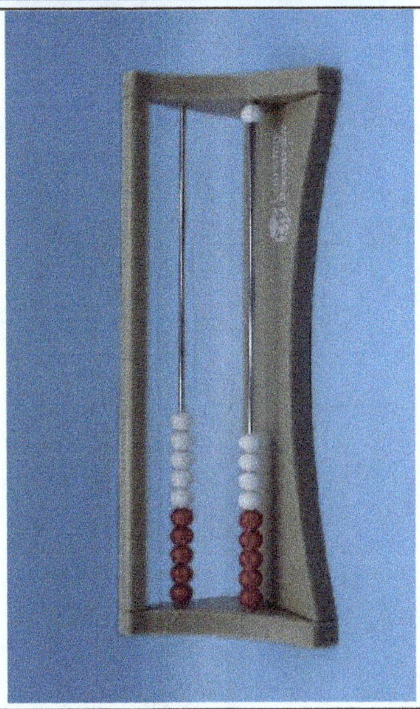

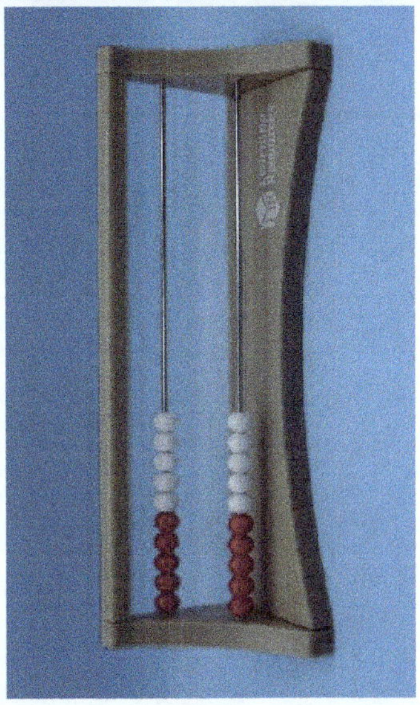

Module 6: Abacus place value subitizing photos

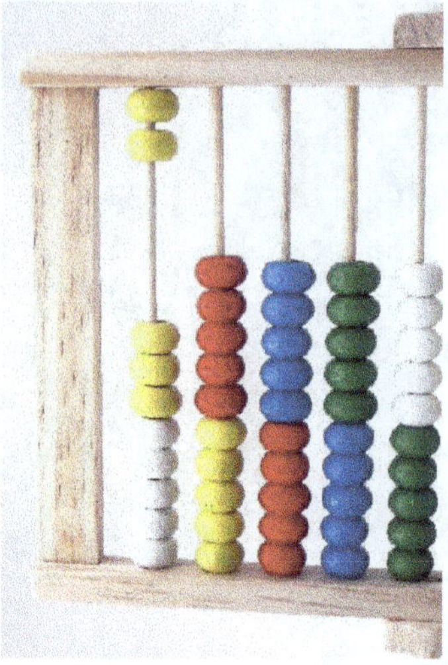

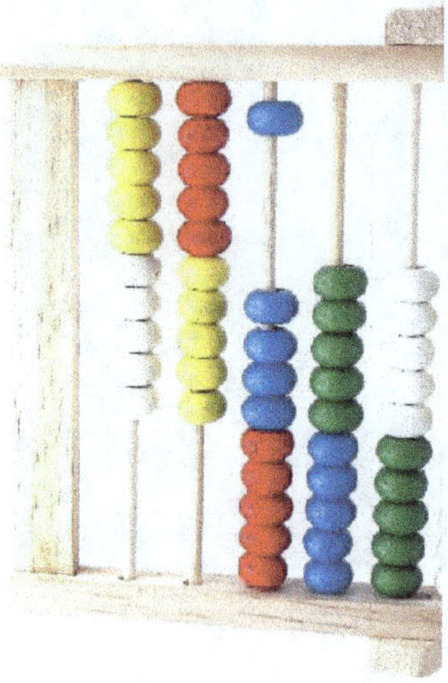

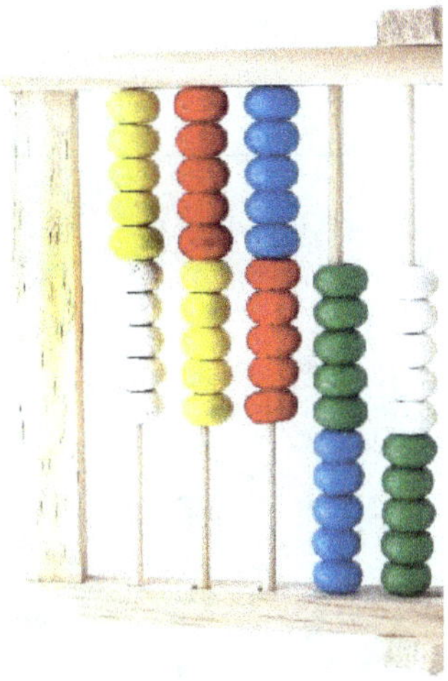

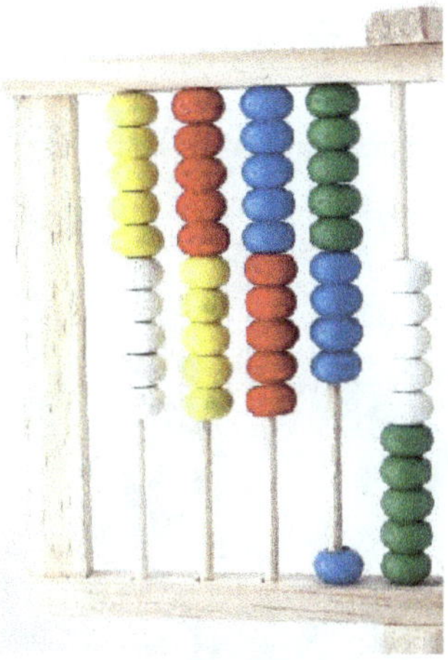

Module 6: Spike abacus

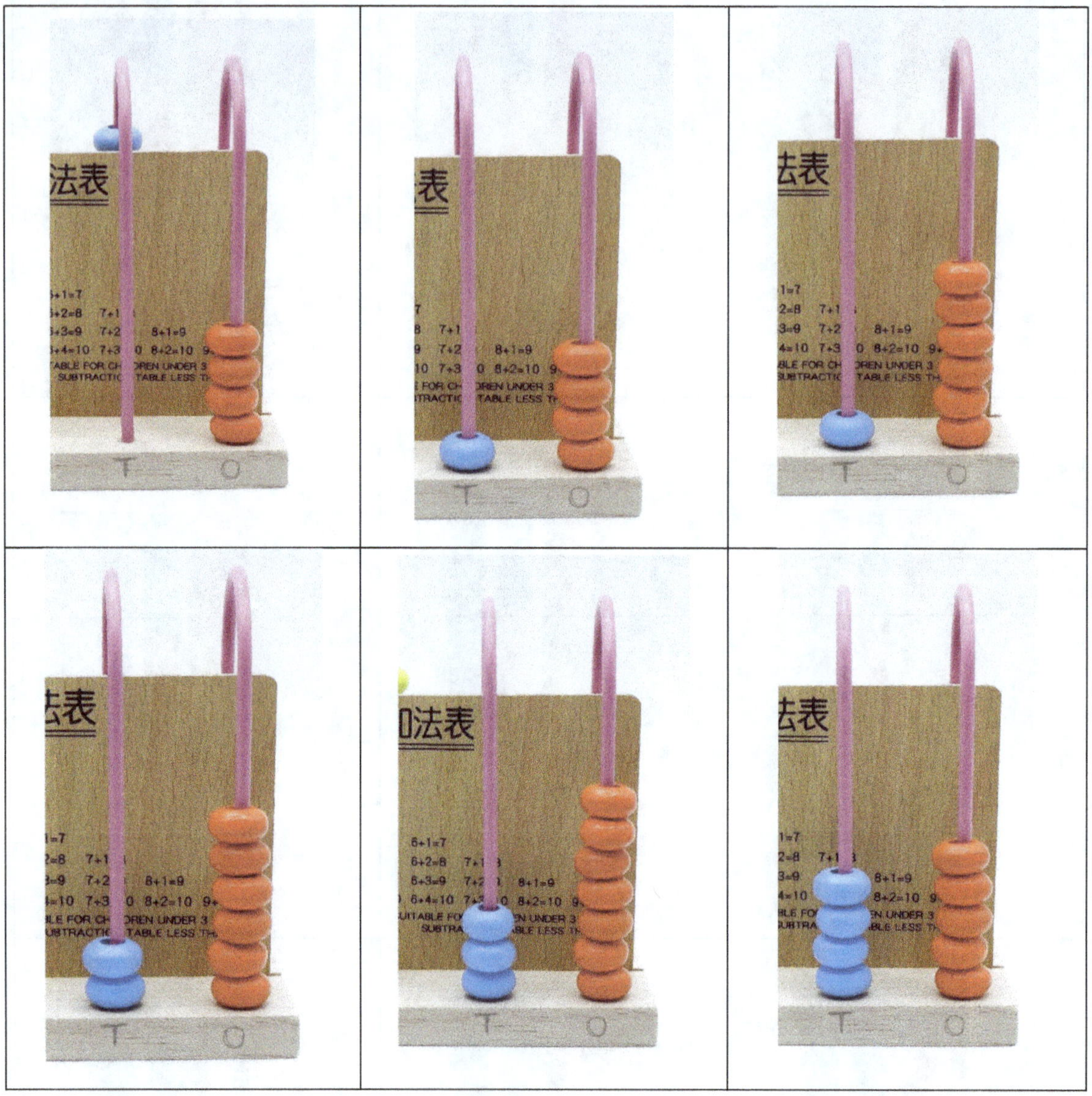

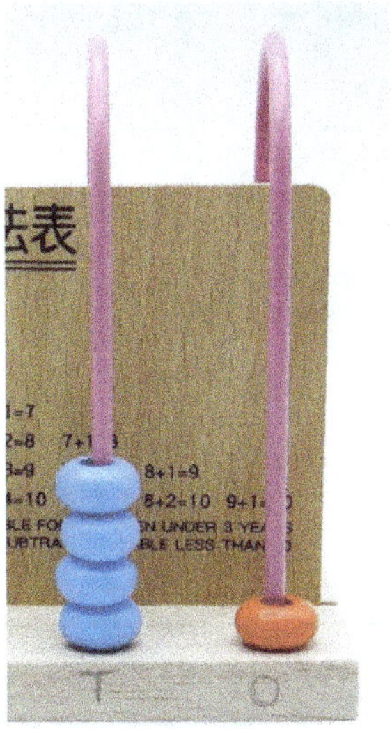

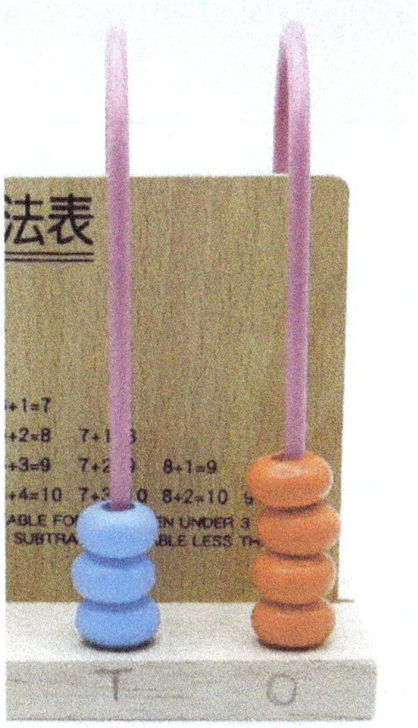

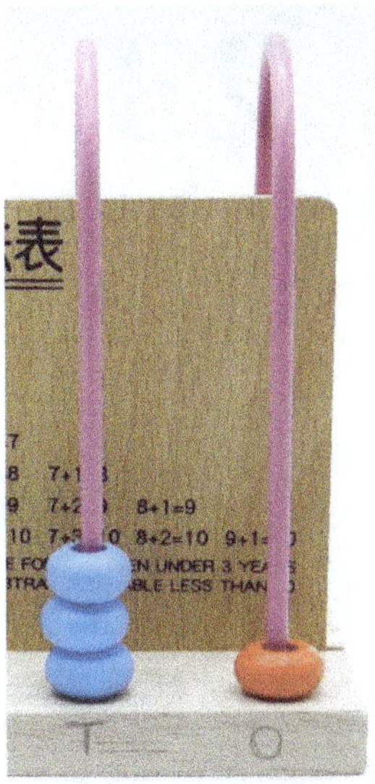

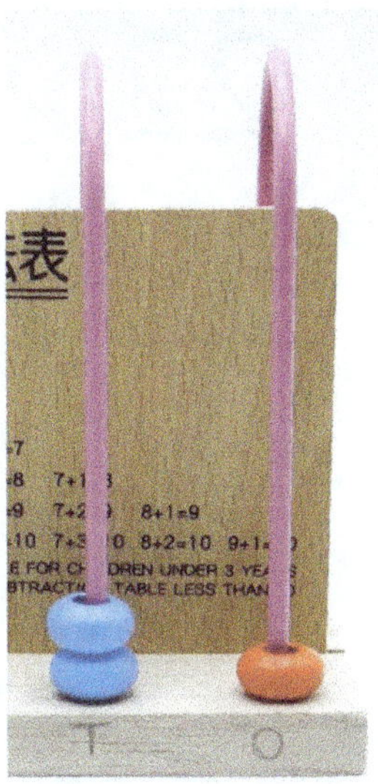

Module 6: Go fish, memory, snap, stealing game - find a near double (and solve it)

24	25	23
24	22	23
21	22	20
21	19	20

© Copyright 2024 Mathtastic: Tracy Ashbridge. All rights reserved

18	19	17
18	16	17
15	16	14
15	13	14

12	13	11
12	10	11

Module 7 - Subitizing cards – all cards - photos

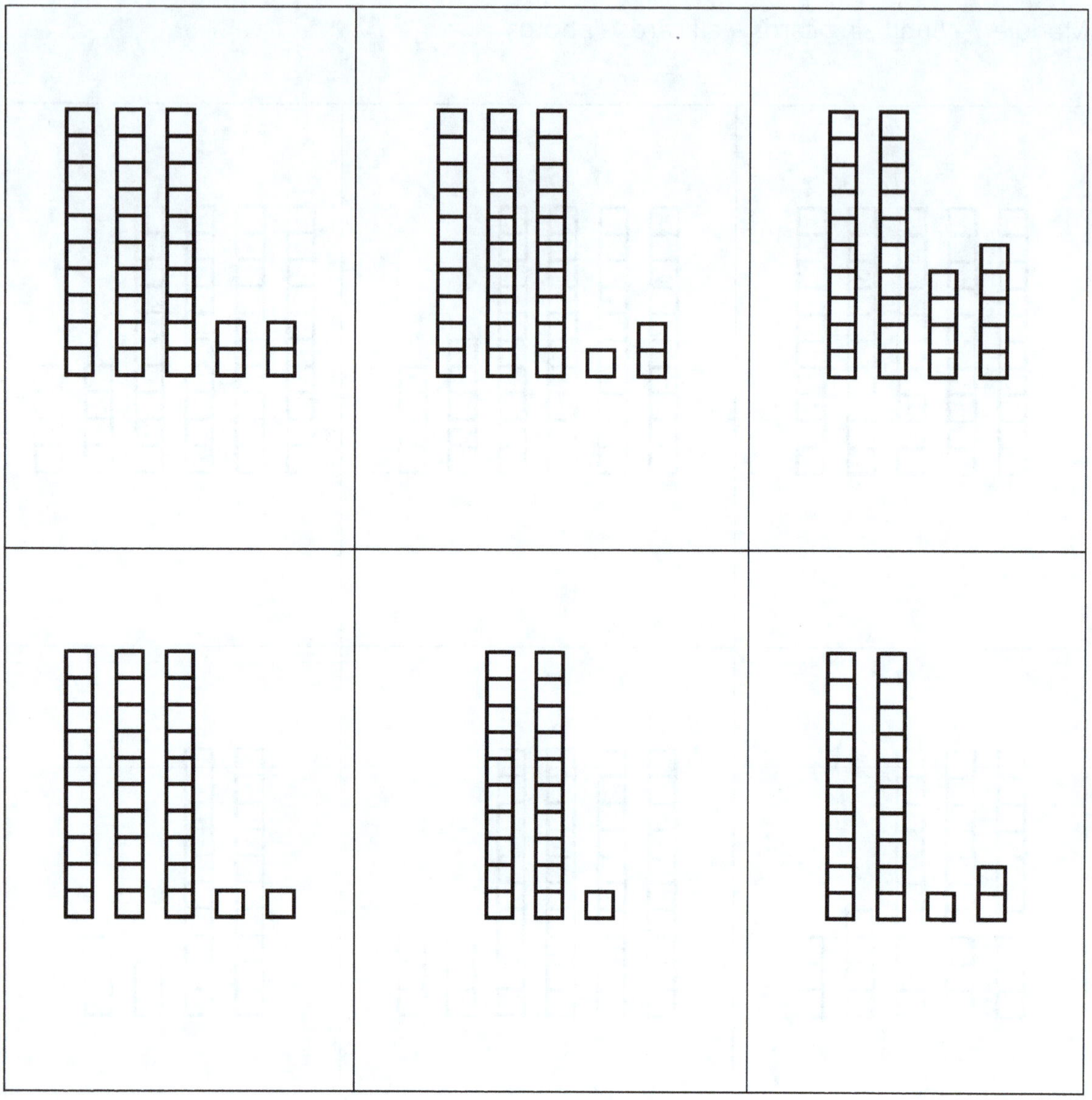

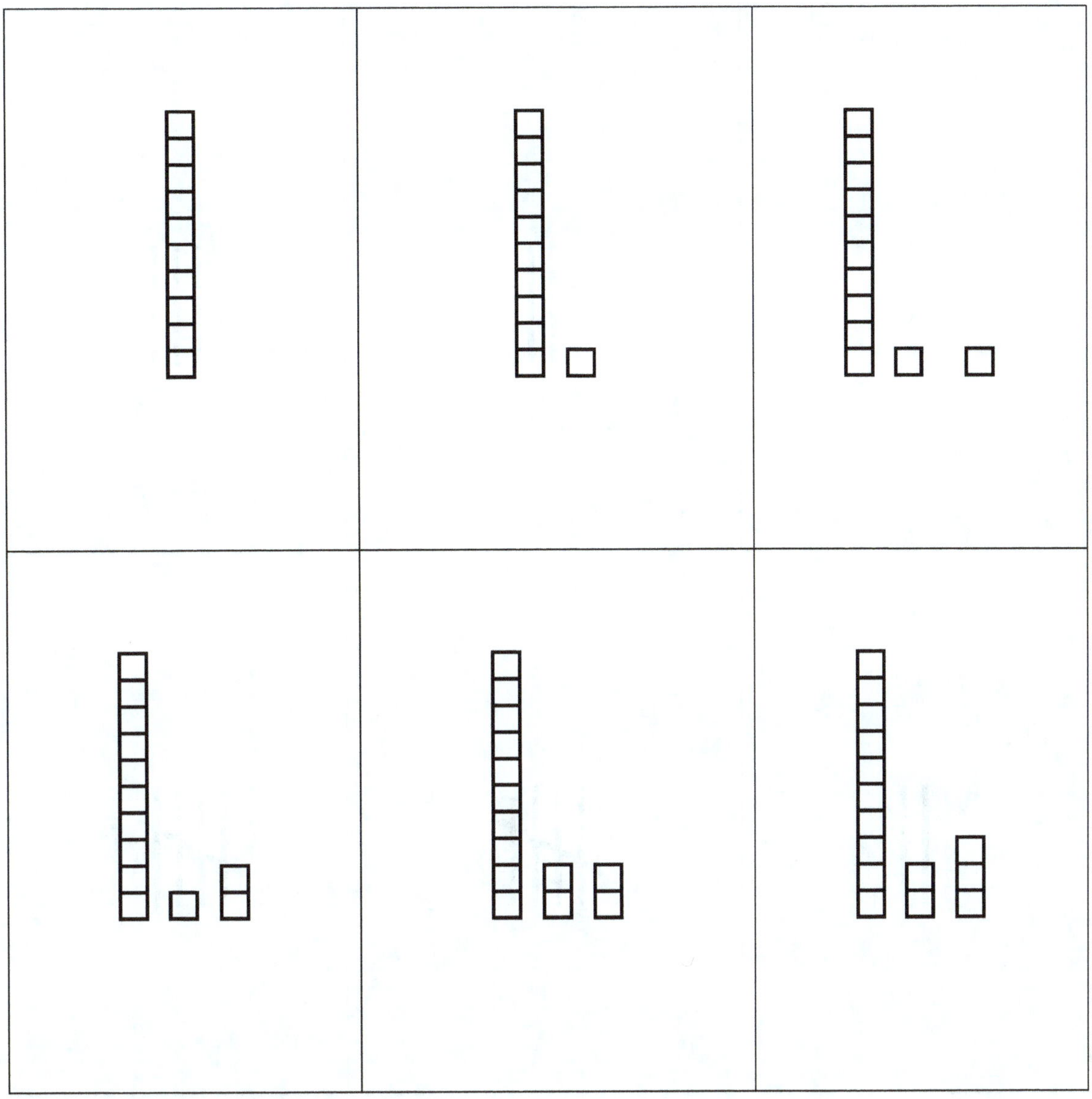

I	II	III
IIII	IIII̷	IIII̷ I

ⅢⅠⅠ	ⅢⅠⅠⅠ	ⅢⅠⅠⅠⅠ
Ⅲ Ⅲ	Ⅲ ⅢⅠ	Ⅲ ⅢⅠⅠ

13	14	13
14	17	18

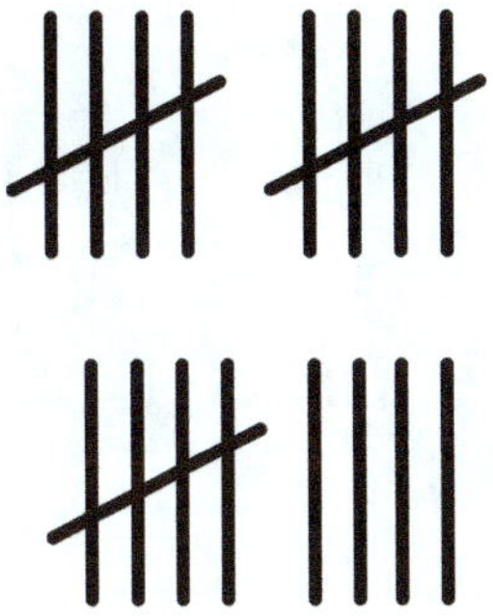

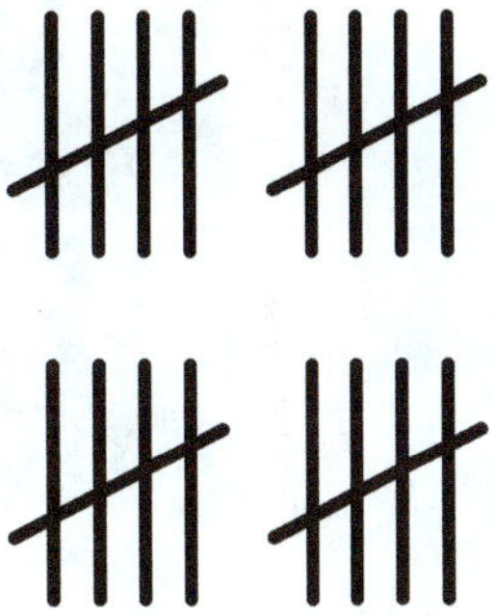

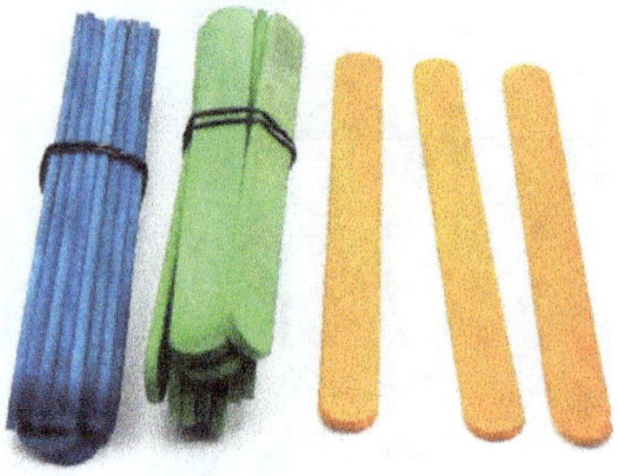

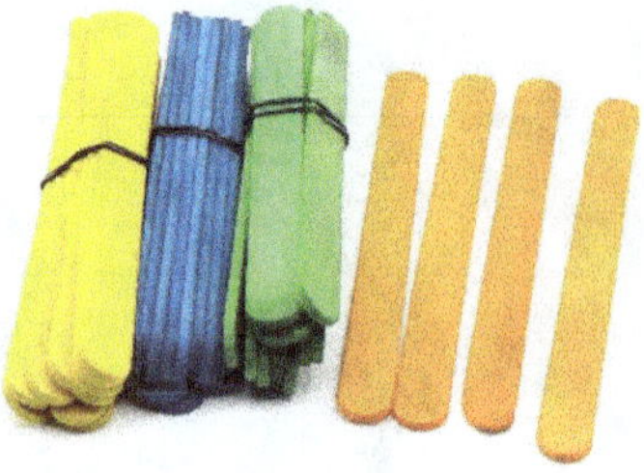

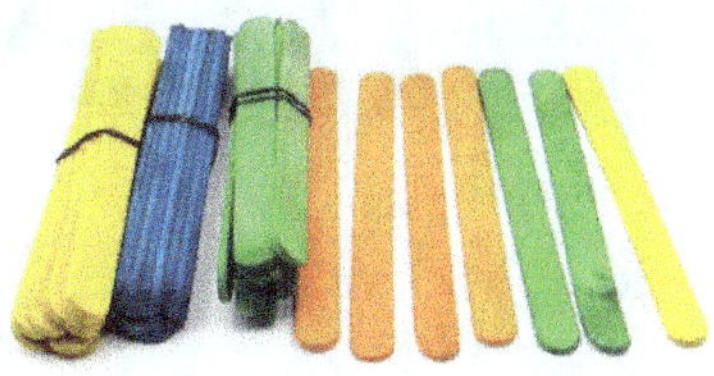

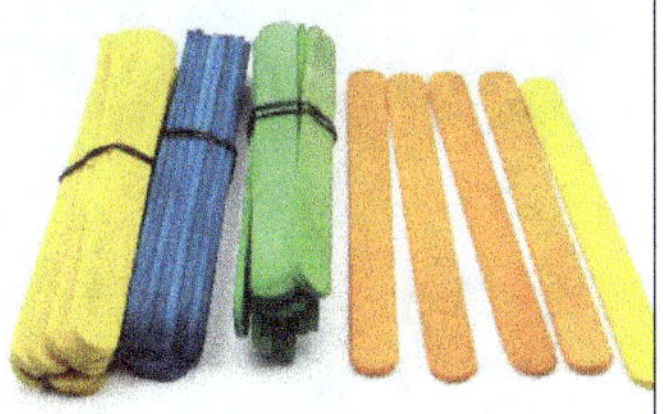

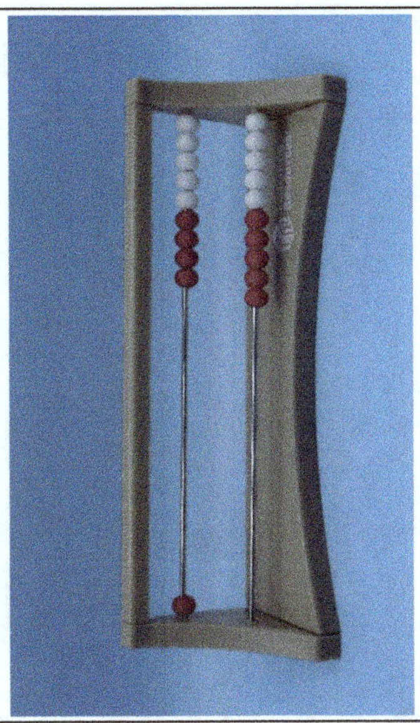

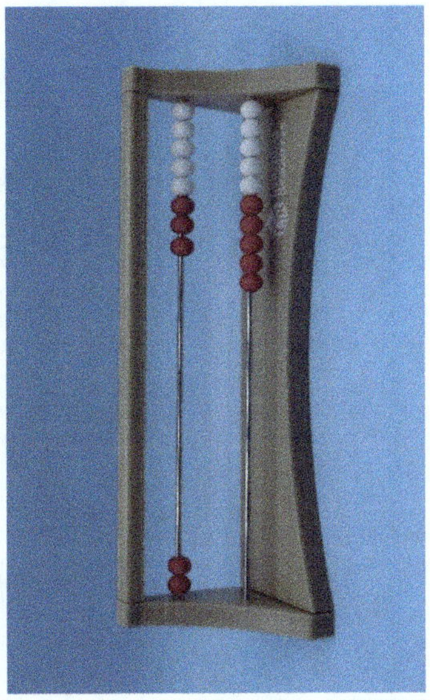

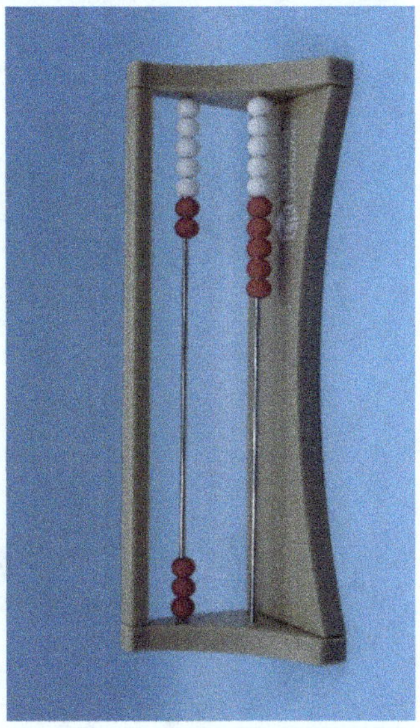

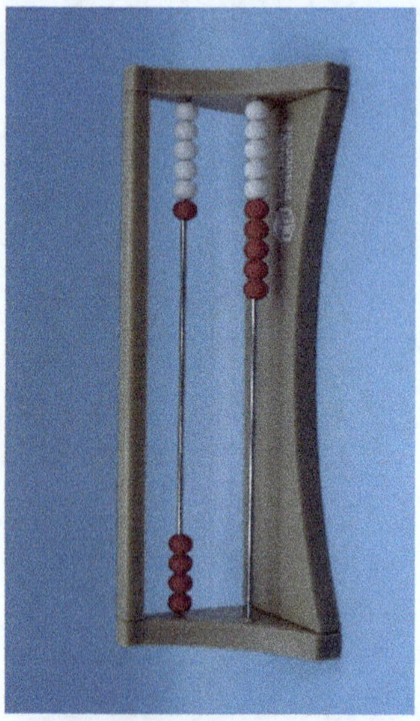

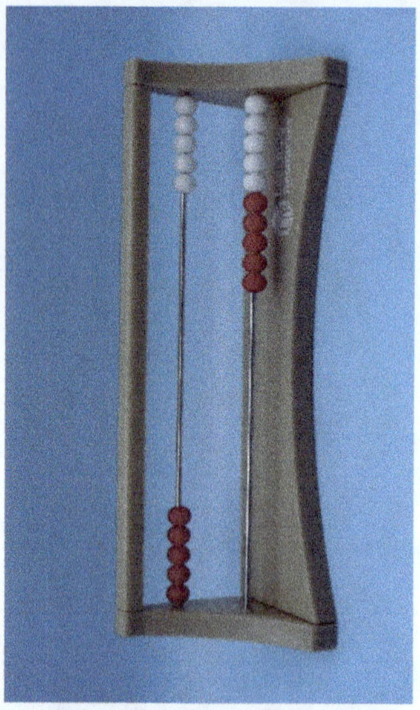

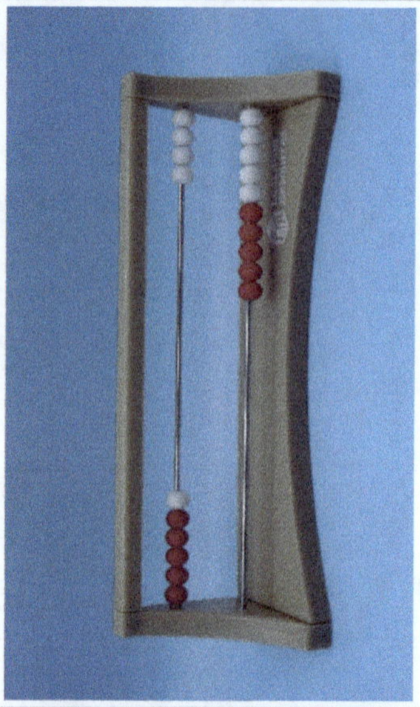

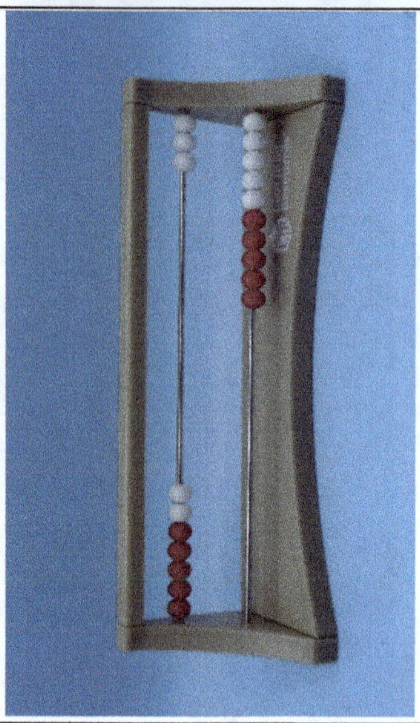

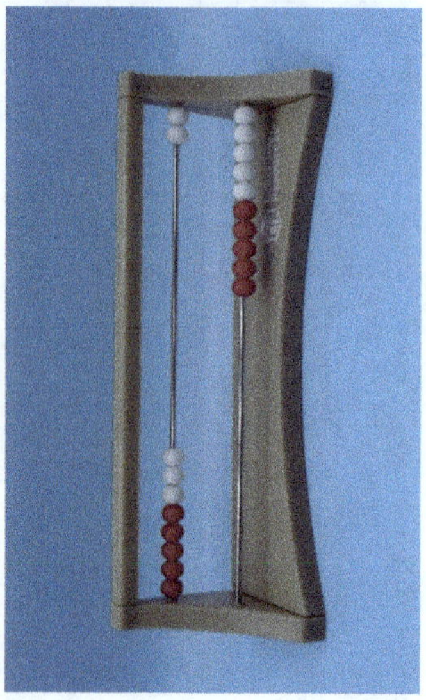

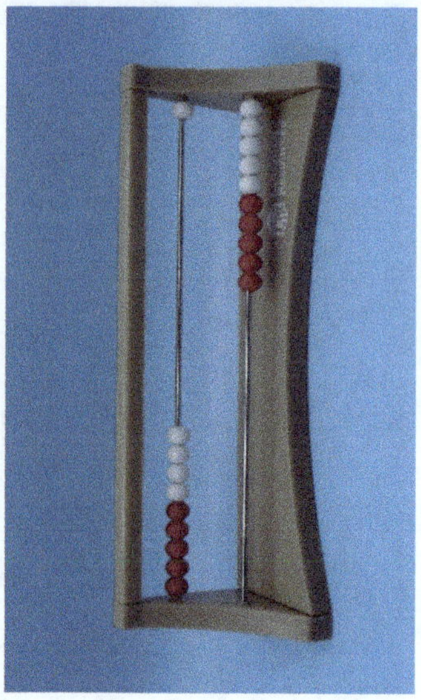

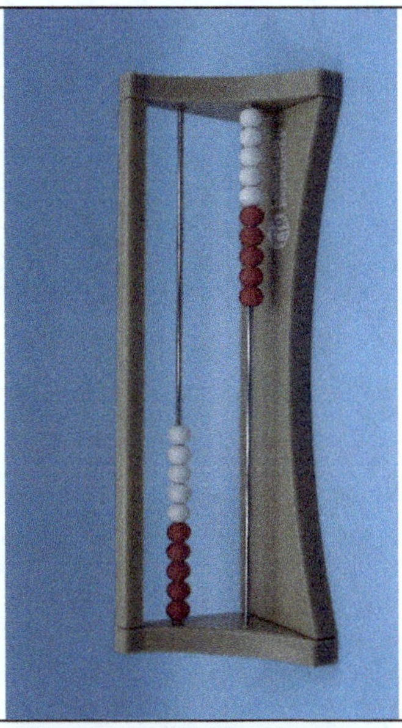

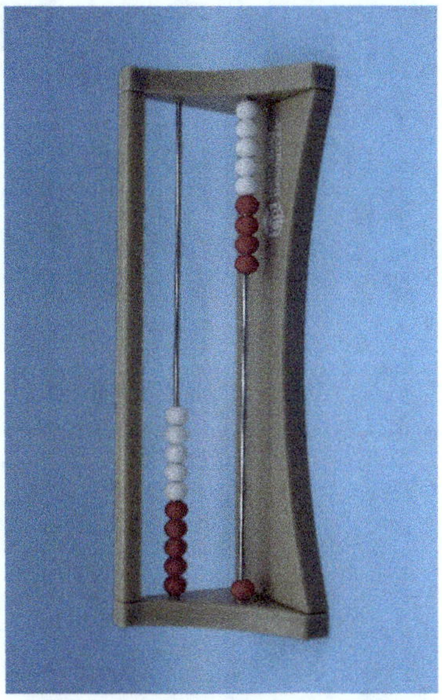

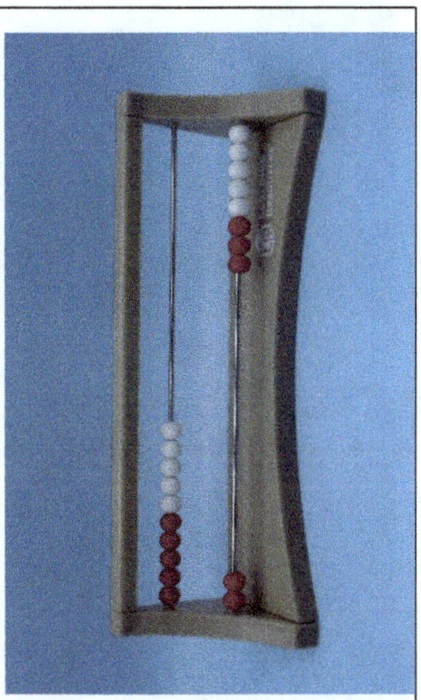

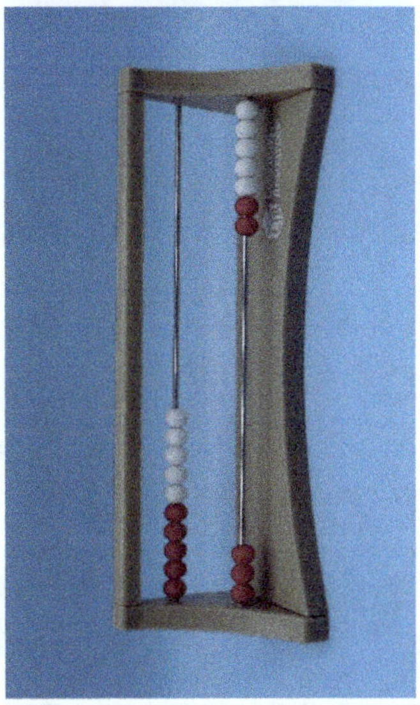

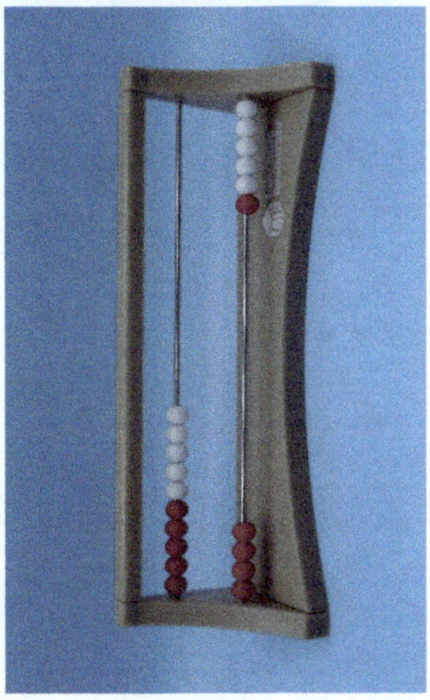

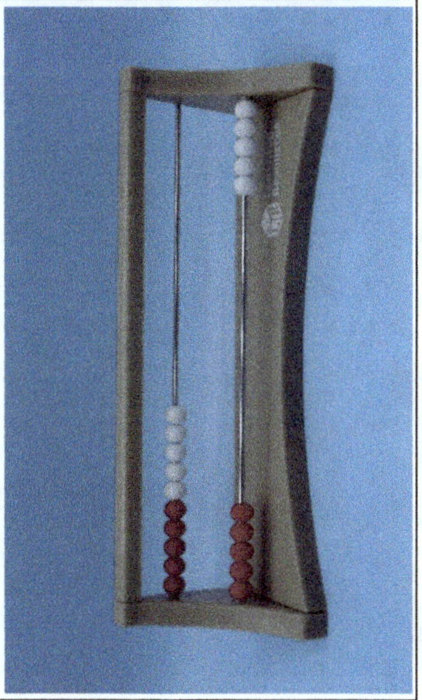

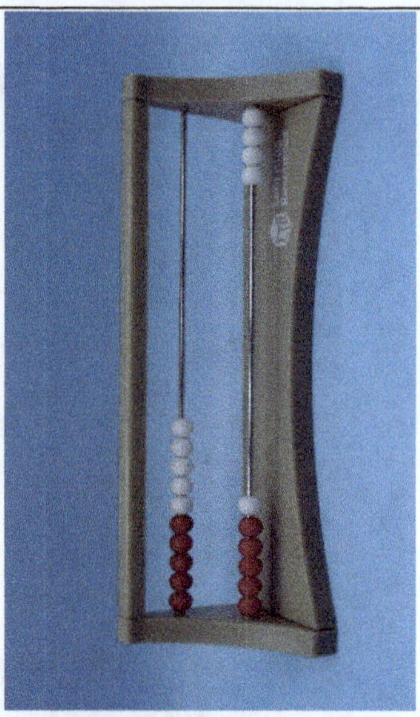

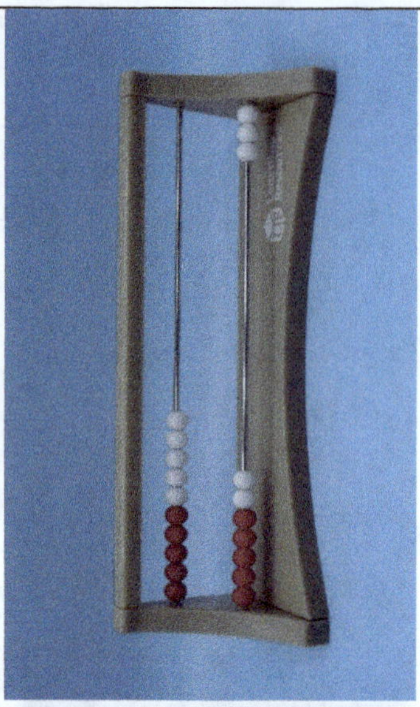

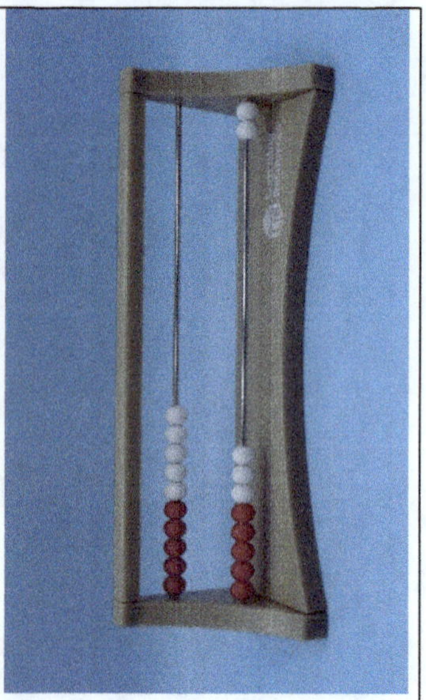

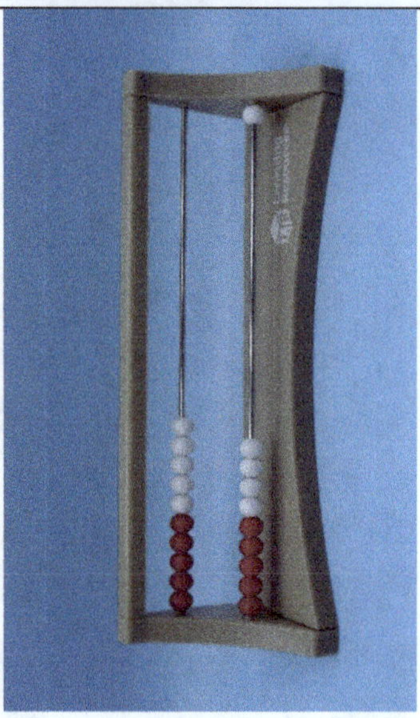

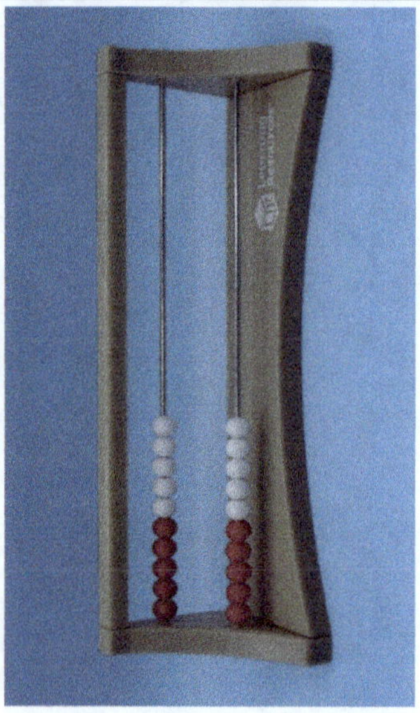

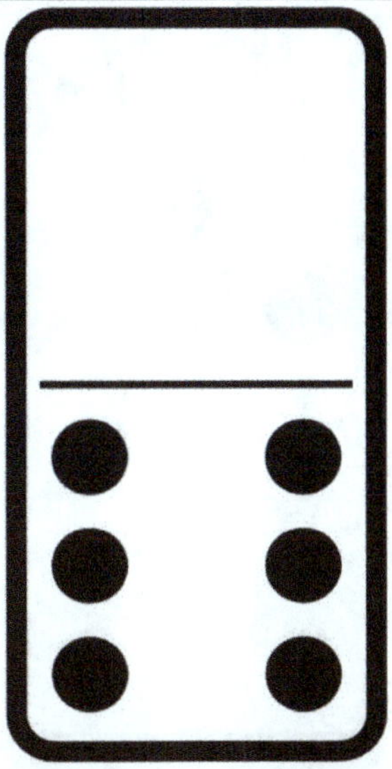

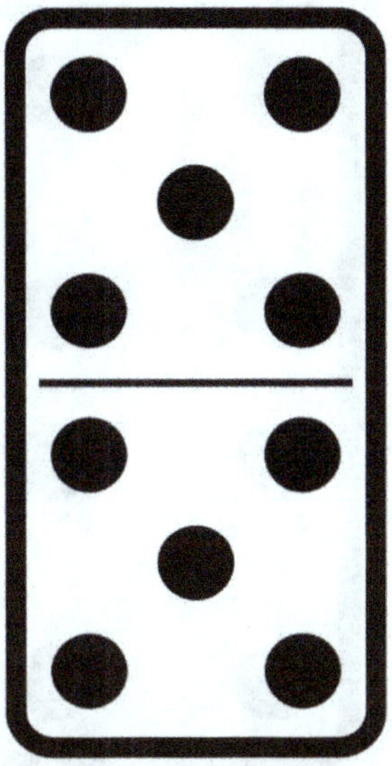

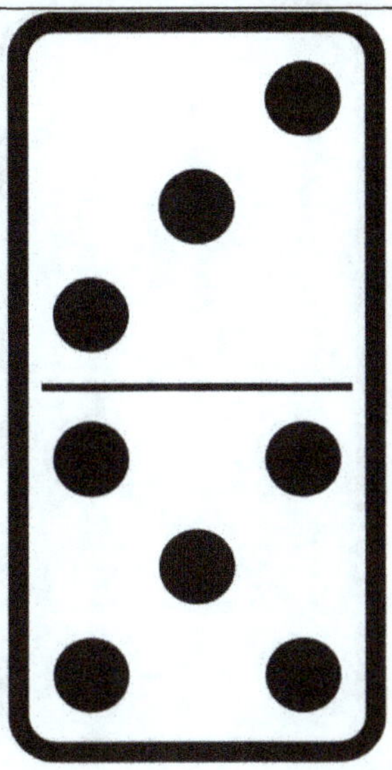

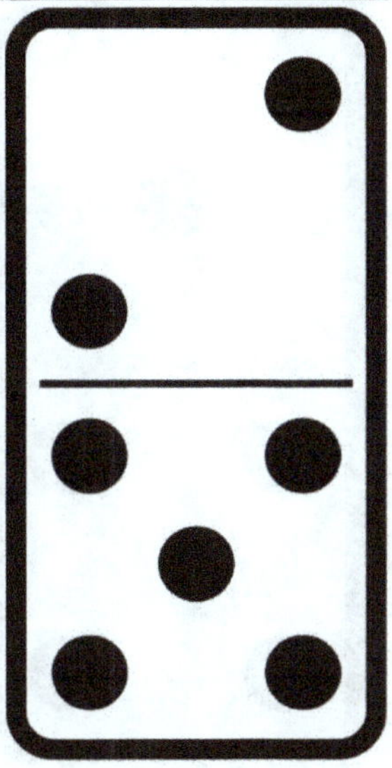

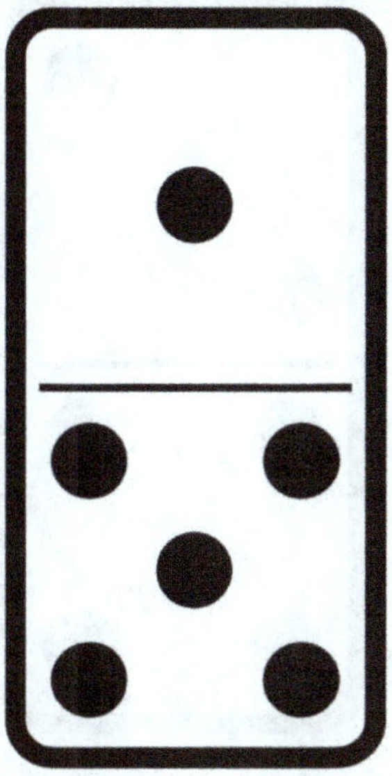

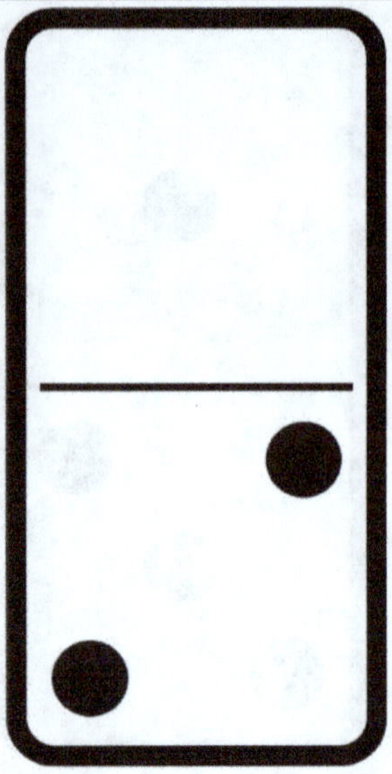

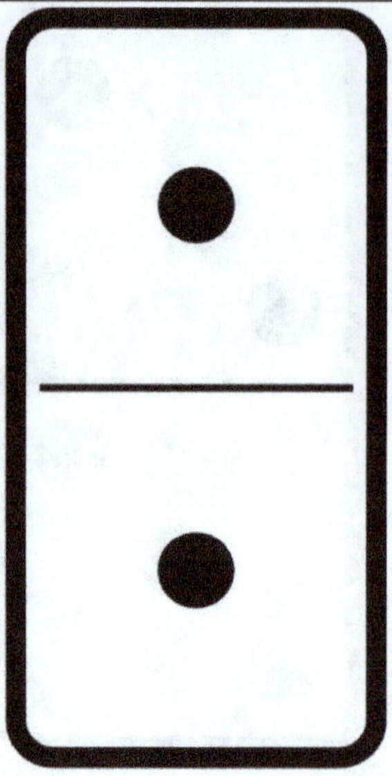

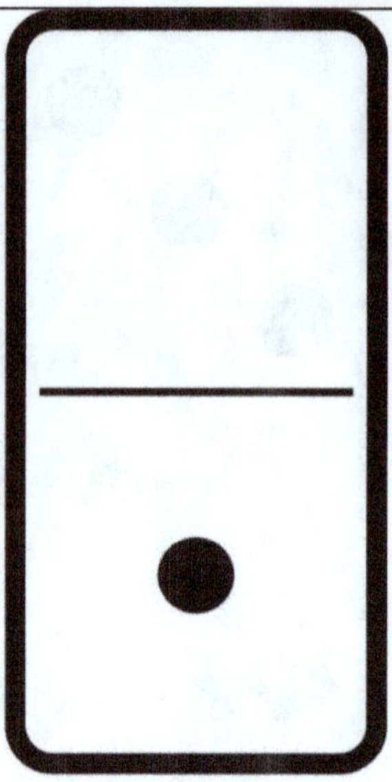

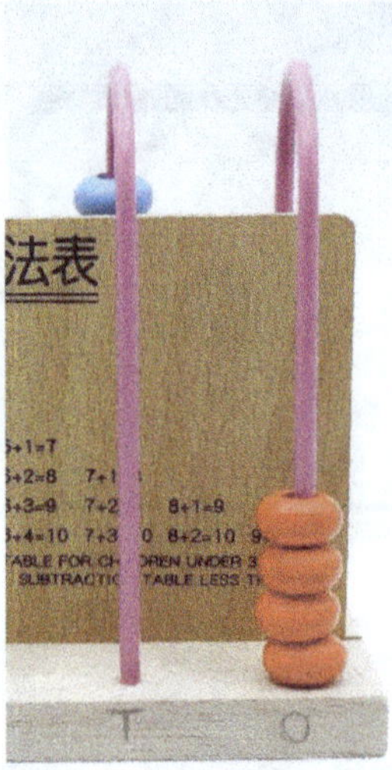

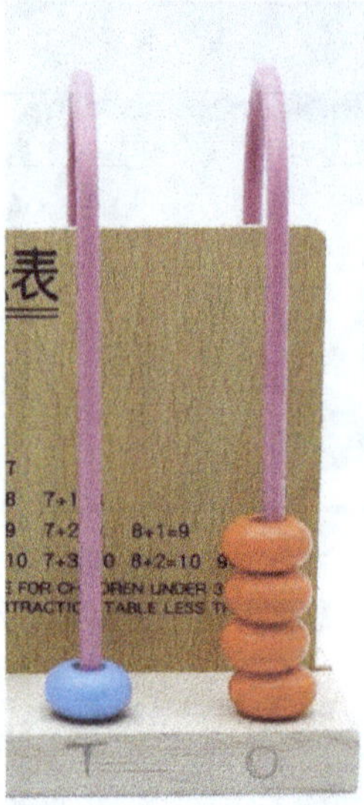

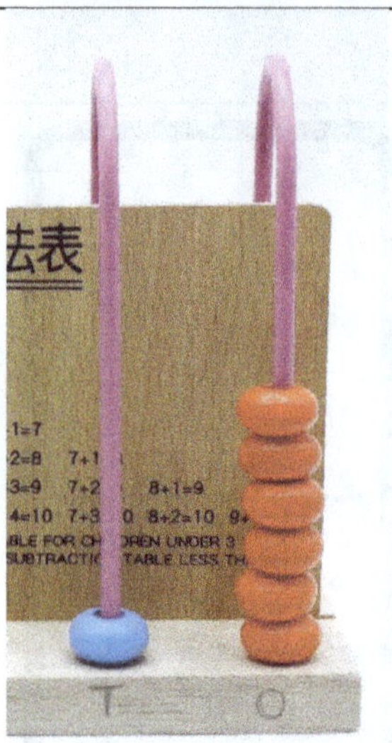

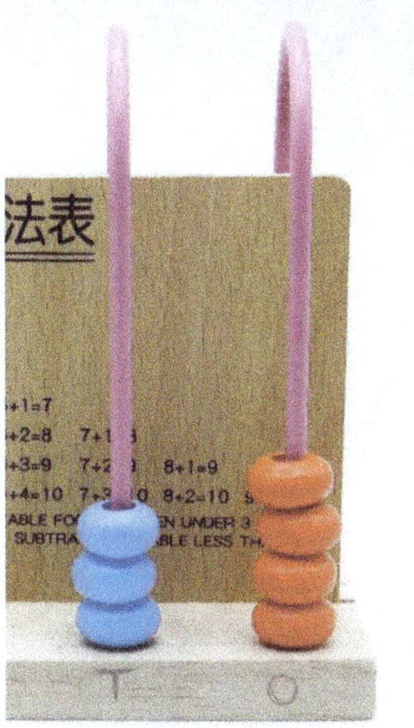

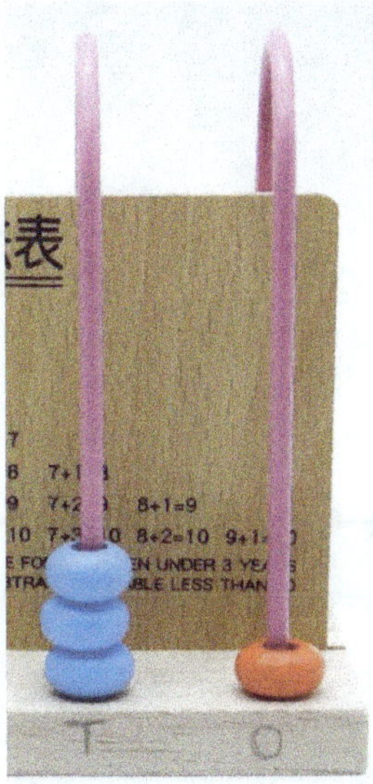

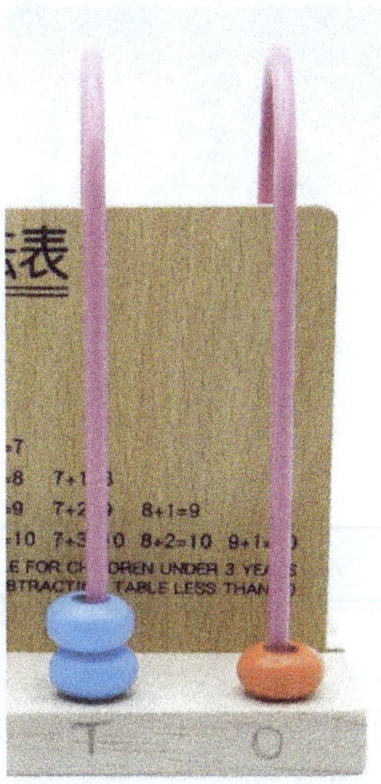

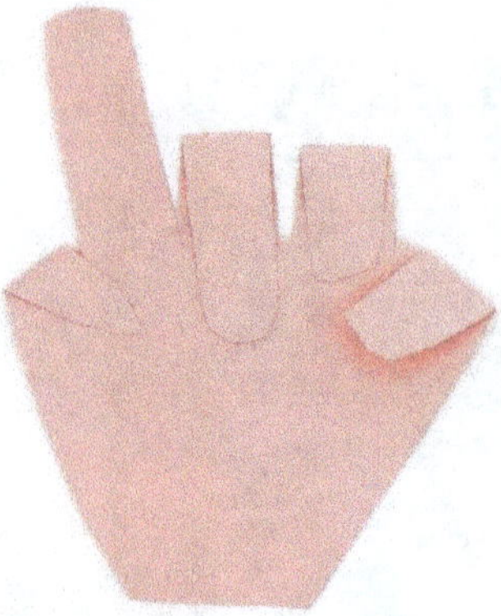

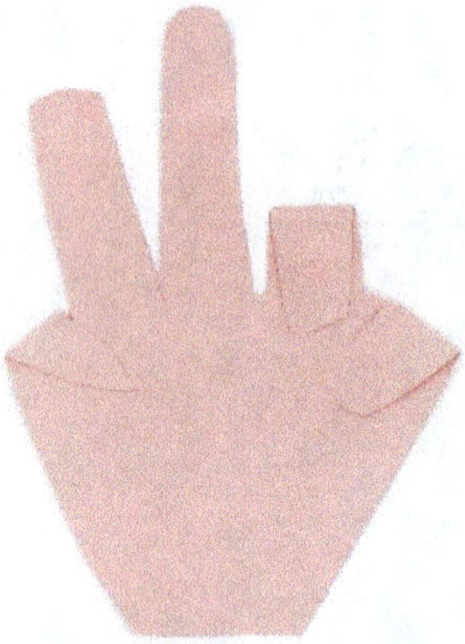

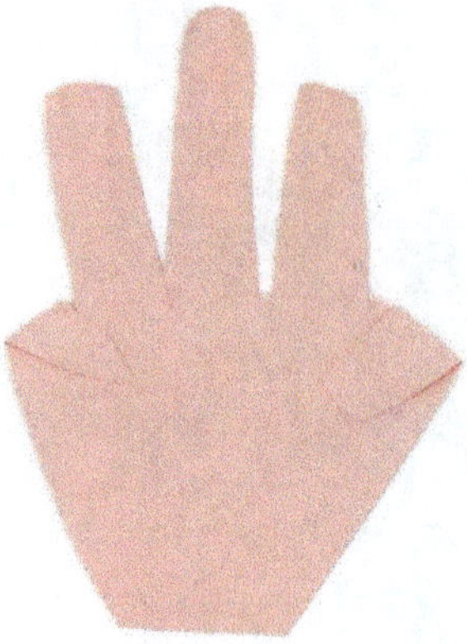

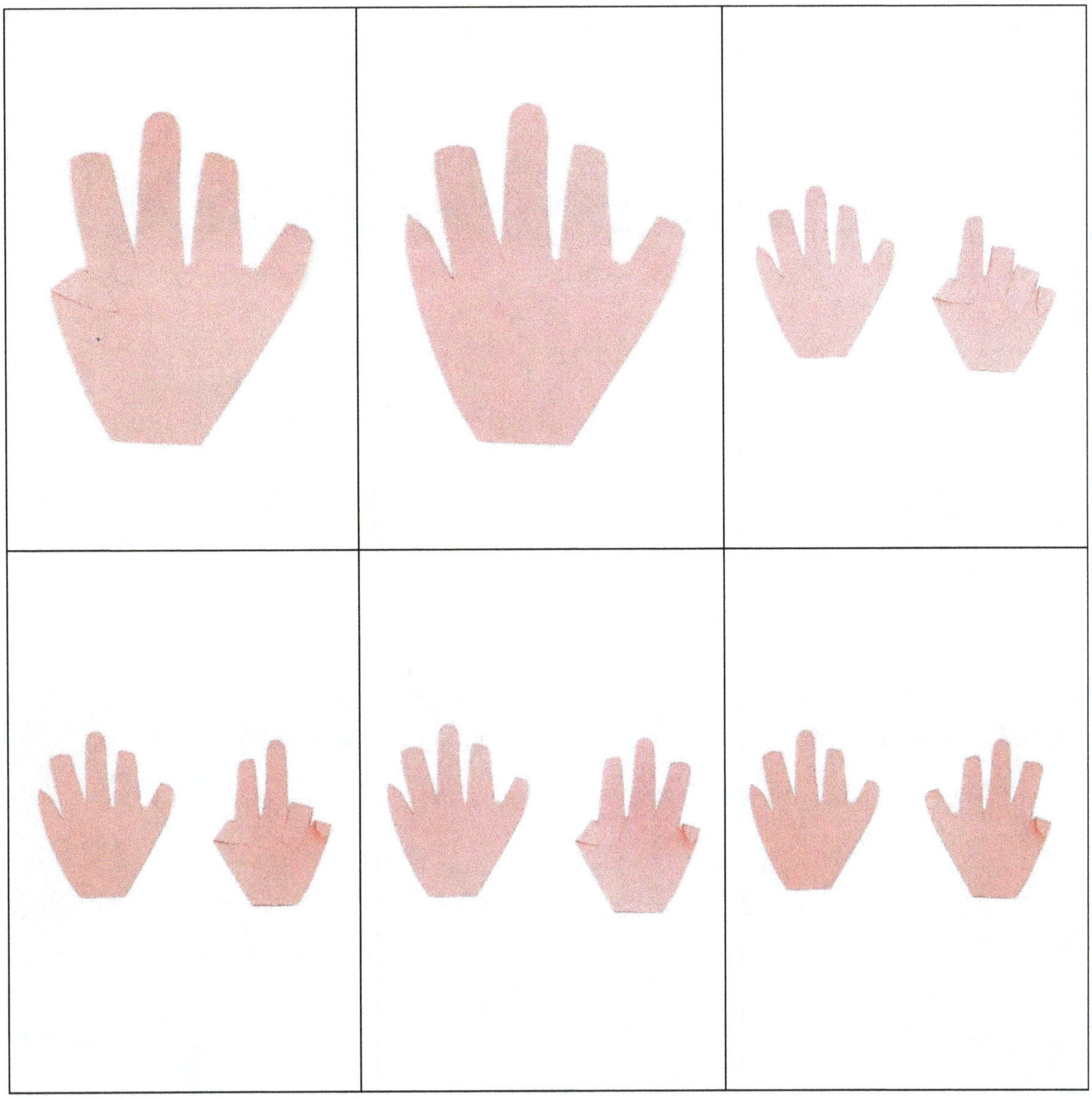

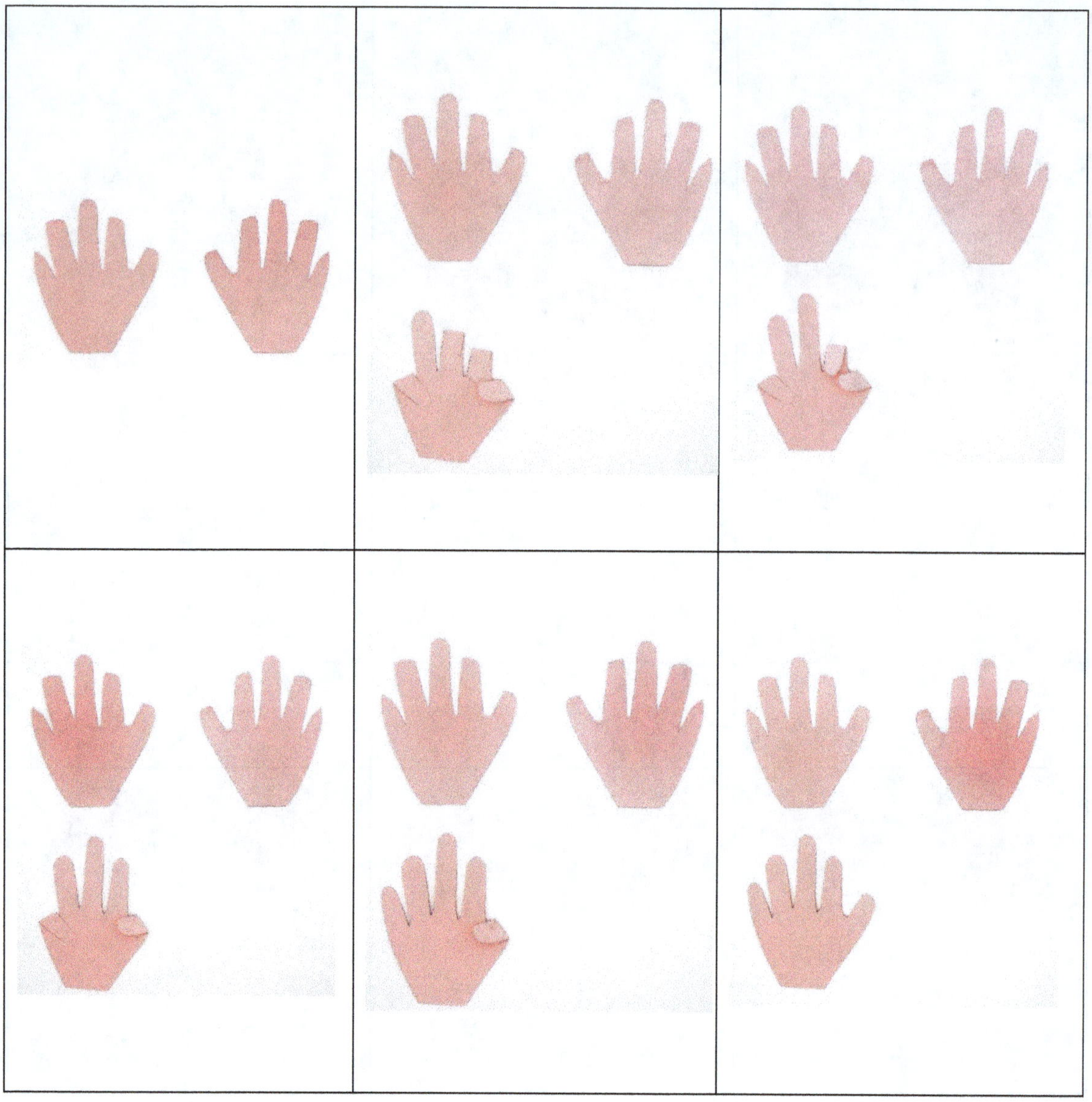

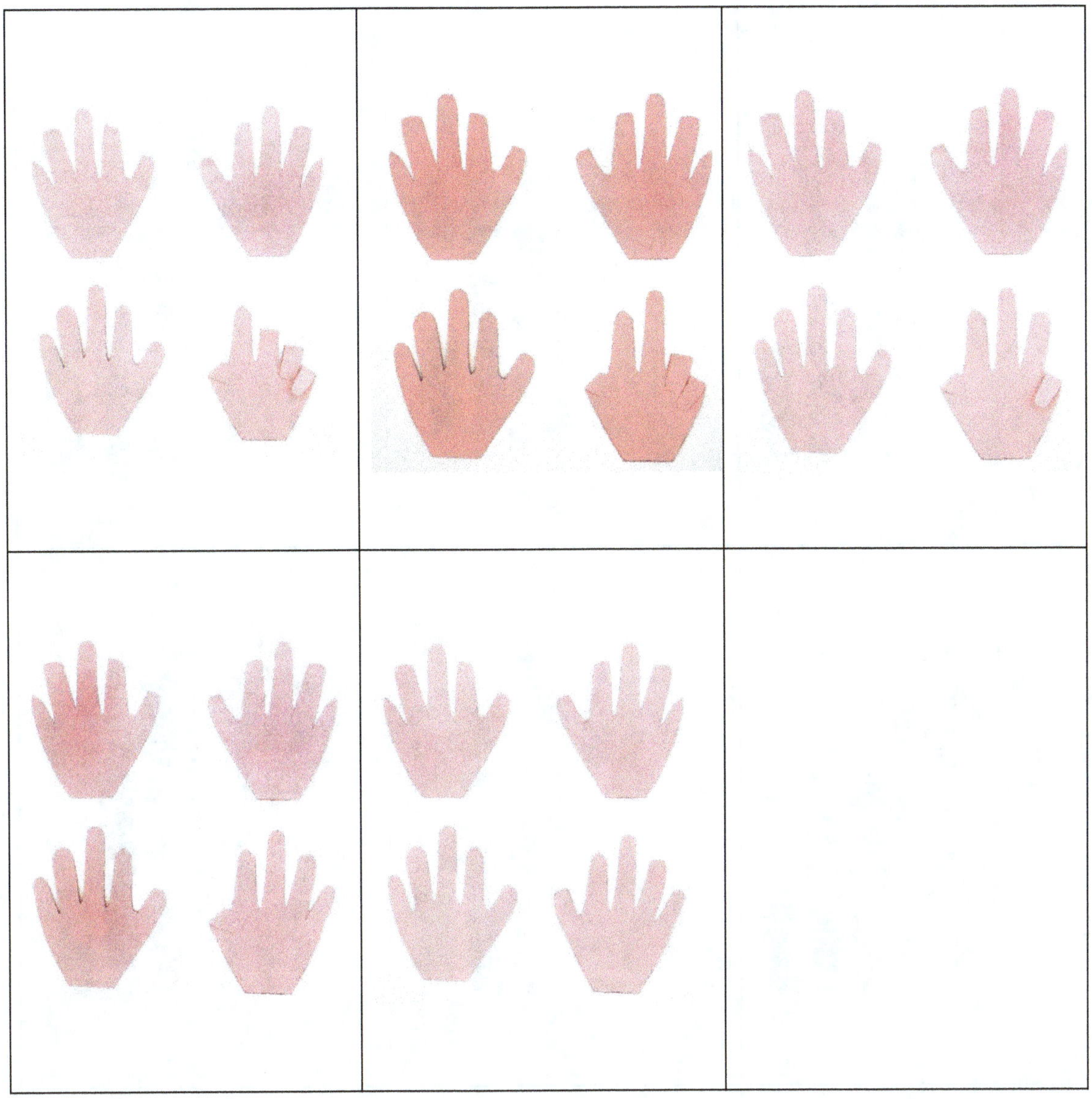

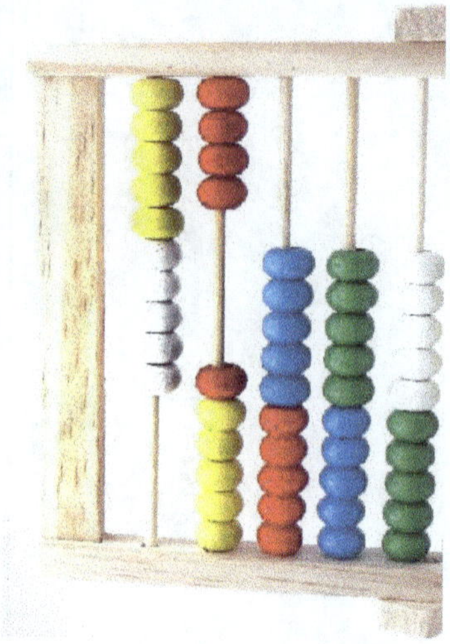

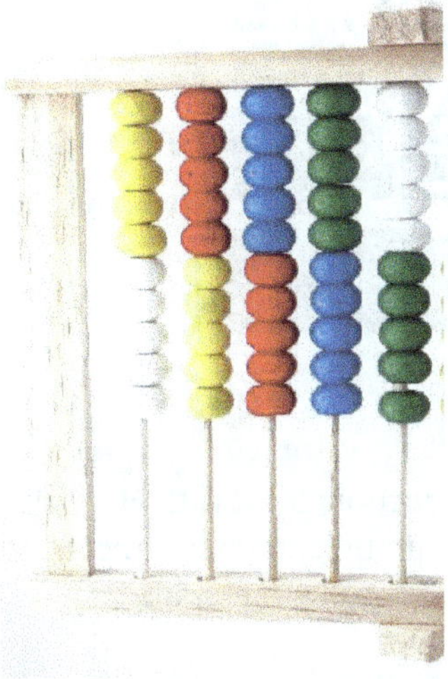

Module 7: Round About game

Round About Place Value Game

This game teaches students to group and regroup for place value. It can be played as either addition or subtraction.

How to play

Print the board game on A3 and collect your place value materials.
Students select a place to start on the track.
Addition: Roll the dice and collect the number you land on using place value materials. As you collect more you need to trade/ regroup ones for tens. The winner is the first person to collect 50.

Watch the video https://youtu.be/PzNZpF3TUPk

Subtraction: Start with 50 in place value materials. (49 is simpler if needed). As the students land on the numbers, they subtract from their place value materials. The winner is the first person to run out of place value materials.

Watch the video https://youtu.be/nk65xAGkmCQ

© Copyright 2024 Mathtastic: Tracy Ashbridge. All rights reserved

Module 7: Go fish, memory, snap, stealing game - find the same number partitioned in another way.

50	5 tens	50 ones
4 tens and 10 ones	45	4 tens and 5 ones

© Copyright 2024 Mathtastic: Tracy Ashbridge. All rights reserved

3 tens and 15 ones	2 tens and 25 ones	32
3 tens and 2 ones	30 ones	1 ten and 22 ones
28	2 tens and 8 ones	28 ones

1 ten and 18 ones	22	2 tens and 2 ones
22 ones	1 ten and 12 ones	47
4 tens 7 ones	47 ones	3 tens and 17 ones

Level 3

Numbers to 50

Workbook

© Copyright 2024 Mathtastic: Tracy Ashbridge. All rights reserved

Instructions

Answer the questions on each page. Each row is for a different day. The first column contains practice related to the work in the module and the second column is for practice from previous modules. Problems should be presented alternatively horizontally and vertically from Module 2 onwards.

Students (or adults) should copy the problem into a maths exercise book – this can be plain, lined or squared. In the early stages a plain book may be easier.

Divide the page into 2 columns – the first for writing the problem and the second to represent in the 5 different ways. See below for an example.

For each of the questions in the first column solve the problem and note how it was solved in the third column – see below. Then show understanding using one or more of the ways shown on the conceptual understanding pentagon below.

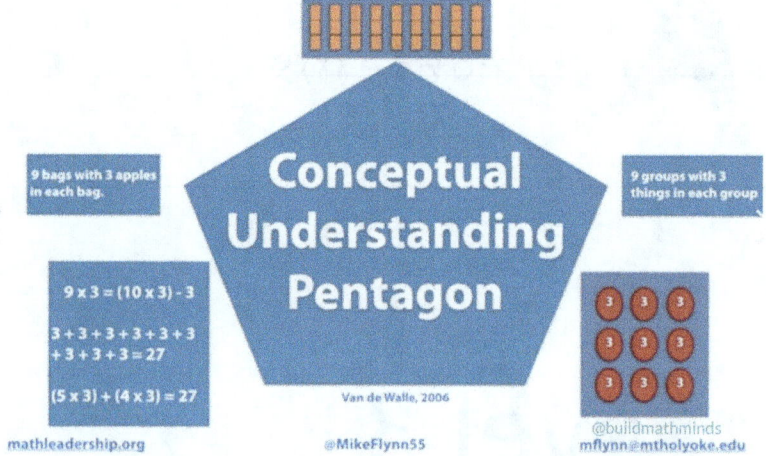

This way students can show their conceptual understanding or misunderstanding:

- Model – show on a model e.g., ten frame, Rekenrek, Numicon
- Words – write out in works – six plus three equals nine.
- Pictures – draw out the problem as a picture.
- Equations – write the equation (already done for you but you could rewrite vertically or horizontally)
- Contexts – write a number story – e.g. I had 3 apples, and I ate 1. Now there are only 2 apples.

© Copyright 2024 Mathtastic: Tracy Ashbridge. All rights reserved

Setting out the student book

$6 + 2 = 8$

Six plus two equals eight.

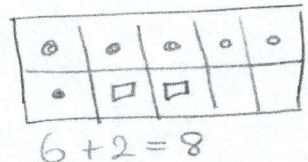

Jane had 6 puppies and her Poppy brought her 2 more. Now she has 8 puppies!

$6 + 2 = 8$

$4 - 1 = 3$

Four subtract one equals three.

Four birds were in the garden. One flew away. Now there are three birds

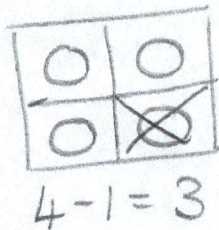

$4 - 1 = 3$

Coding the answers

Coding the answers will help to diagnose which strategies the students are using and which they are not. We really need students to move beyond counting in ones and seeing numbers in bigger groups as well as learning some of these facts so they can recall them automatically.

Coding the student answer strategy: ask the student how they worked out the answer – they may use more than one way	
Automatic – just knew the answer, immediate recall	A
Counted on or back a small number (+/- 0,1,2,3)	S
Rainbow facts – knew the pair made 10	RF
Counted on from largest number	CO
Counted back for subtraction	CB
Doubled or halved	D or H
Near double	ND
Place value – e.g., 10+4= 14	PV
Compensated – made an adjustment to the number before calculating e.g., 9+6 – rearranged to 10+5 as easier	C
Other – you may wish to note this down	O

© Copyright 2024 Mathtastic: Tracy Ashbridge. All rights reserved

Module 1 – Adding and Subtracting 0,1,2,3.

Lesson	Draw and solve	Code	Retrieval Practice – 10 mins per day	Code
1	1. 11+1= 2. 23-2= 3. 3+12= 4. 35-3= 5. 14+?=17 6. 37-?=35 7. 3+?=22 8. ?-3=35 9. ?+19=22 10. 44+2=		11. 19+1= 12. 15+1= 13. 15-2= 14. 15-3= 15. 13+6= 16. 17-5= 17. 15+?=20 18. 8+10= 19. 7+7= 20. 7+8= 21. 13-3= 22. 9+4=	
2	1. 2+11= 2. 27-1= 3. 13+?=14 4. 48-2= 5. ?+15=16 6. 45-?=43 7. 42+1= 8. ?-2=42 9. 43+?=45 10. ?-1=33		11. 15+1= 12. 12+1= 13. 11-2= 14. 13-3= 15. 17-1= 16. 17+3= 17. 2+16= 18. 13-1= 19. 14-?=12 20. ?-3=14	

© Copyright 2024 Mathtastic: Tracy Ashbridge. All rights reserved

3	1. 11+?=13 2. 23-2= 3. 2+?=13 4. 26-3= 5. 16+2= 6. ?-2=26 7. 42+2= 8. 36-?=33 9. 44+?=45 10. 46-3=		11. 15-1= 12. 13+3= 13. 3+13= 14. 15-2= 15. 8+9= 16. 7+8= 17. 6+7= 18. 10+5= 19. 4+10= 20. 16-10= 21. 12-?=10 22. 9+4= 23. 13-4=	
4	1. ?+12=13 2. 13-2= 3. 14+1= 4. 17-3= 5. 3+16= 6. 19-?=16 7. 42+?=45 8. ?-2=18 9. ?+46=48 10. 19-1=		11. 19-1= 12. 16+3= 13. 3+16= 14. 13-2= 15. 12+6= 16. 14-5= 17. 7+?=20 18. 14+?=20 19. 7+7= 20. 18-9=	

| 5 | 1. 12+2=
2. 14-2=
3. 2+14=
4. 35-1=
5. 17+?=19
6. ?-3=46
7. ?+43=44
8. 44-2=
9. 3+?=47
10. 25-?=22 | | 11. 19-1=
12. 16+3=
13. 3+16=
14. 13-2=
15. 7+?=20
16. 5+6=
17. 15-7=
18. 18-10=
19. 10+?=16
20. 14-5=
21. 13+5= | |

Module 2 – Add from largest, Subtract by counting back.

Lesson	Draw and solve	Code	Retrieval Practice – 10 mins per day	Code
1	1. 31+5= 2. 37+4= 3. 36+5= 4. 42+5= 5. 22+4= 6. 32-5= 7. 39-4= 8. 41-6= 9. 22-4= 10. 26-4=		11. 32+1= 12. 45-3= 13. 13+7= 14. 20-16= 15. 14-4= 16. ?+5=15 17. ?+6=12 18. 6+7= 19. 2+10= 20. ?+7=17 21. 9+4= 22. 12-4=	
2	1. 33+5= 2. 38-5= 3. 25+6= 4. 26-5= 5. 27+4= 6. 43-6= 7. 31+?=37 8. 37-?=32 9. ?+5=44 10. ?-4=17		11. 28+5= 12. 31-5= 13. 39+5= 14. 39+6= 15. 42-5= 16. 21-5= 17. 36-?=32 18. ?-7=28 19. 23-?=19 20. ?-3 =29	

© Copyright 2024 Mathtastic: Tracy Ashbridge. All rights reserved

3	1. 31+6=		11. 34+3=	
	2. 38-4=		12. 45-2=	
	3. 21+5=		13. ?-3=46	
	4. 28-4=		14. 38+?= 41	
	5. 44+5=		15. 14+6=	
	6. 23-4=		16. 20-12=	
	7. 45-?=41		17. 14-10=	
	8. 29+?=33		18. ?-10=8	
	9. ?+4=39		19. 8+4=	
	10. ?-7=25		20. 13-4=	
4	1. 28+4=		11. 4+10=	
	2. 32-5=		12. 17-7=	
	3. 38+6=		13. ?-8=10	
	4. 37+6=		14. ?-4=10	
	5. 44-5=		15. 5+6=	
	6. 27-5=		16. 7+?=15	
	7. 38-?=32		17. 7+7=	
	8. ?-7=26		18. 9+9=	
	9. 25-?=19		19. 17-8=	
	10. ?-5=29		20. 16-8=	
5	1. 28+6=		11. 43+3=	
	2. 44+4=		12. 32-3=	
	3. 27-5=		13. 39+1=	
	4. 32-4=		14. 46+2=	
	5. 38+5=		15. ?+3=35	
	6. 39+6=		16. 14-10=	
	7. 43-5=		17. 17-7=	
	8. 32+5=		18. ?-10=10	
	9. 38-6=		19. 37+5=	
	10. 27+6=		20. 42-5=	

© Copyright 2024 Mathtastic: Tracy Ashbridge. All rights reserved

Module 3 – Rainbow Facts

Lesson	Draw and solve	Code	Retrieval Practice – 10 mins per day	Code
1	1. 23+7= 2. 45+5= 3. 34+6= 4. 28+?=30 5. 17+?=20 6. 42+?= 7. ?+35=40 8. 19+?=20 9. 11+39= 10. 33+17=		11. 25+3= 12. 37-1= 13. 32-4= 14. 36+?=40 15. 6+10= 16. 9+?=13 17. 45+?=47 18. ?-2=29 19. 37+5= 20. ?+22= 21. 17-?=10 22. 7+7= 23. 8+8= 24. 8+9=	
2	1. 12+8= 2. 41+?= 3. 34+6= 4. 15+35= 5. 32+18= 6. ?+27=30 7. 38+?=40 8. ?-8=22 9. ?-9=41 10. ?-38=2		11. 36+4= 12. 28+2= 13. 44+6= 14. 12+?=30 15. 16+?=20 16. 22+18= 17. 34+?=50 18. 33+17= 19. 13+27= 20. 29+21=	

© Copyright 2024 Mathtastic: Tracy Ashbridge. All rights reserved

3	1. 21+9=		11. 42-2=	
	2. 42+8=		12. 7+36=	
	3. 37+3=		13. 21+?=30	
	4. 25+?=30		14. 19+4=	
	5. 12+?=20		15. 27-?=26	
	6. 35+15=		16. 36+?=39	
	7. 48+?=50		17. 29-6=	
	8. 14+16=		18. 40-?=38	
	9. 24+16=		19. 14-5=	
	10. 19+21=		20. ?+3=50	
4	1. 26+4=		11. 36+?=40	
	2. 48+2=		12. ? + 25 = 30	
	3. 34+6=		13. 10 + 7 =	
	4. 22+?=30		14. ? + 10 = 15	
	5. 18+?=20		15. 6 + 6 =	
	6. 32+18=		16. 9 + 4 +	
	7. 44+?=50		17. ? + 7 = 50	
	8. 13+17=		18. ? + 8 = 18	
	9. 33+17=		19. 9 + 8 =	
	10. 19+21=		20. 5 + 8 =	
5	1. 25+5=		11. 43 + 2 =	
	2. 41+9=		12. 38 + ? = 42	
	3. 34+6=		13. 37 + ? = 40	
	4. 24+?=30		14. 37 + ? = 50	
	5. 12+?=20		15. 9 + 6 =	
	6. 46+?=		16. 38 - 3 =	
	7. ?+25=40		17. ? + 6 = 33	
	8. 29+?=40		18. 28 + ? = 30	
	9. 15+35=		19. 36 + ? = 40	
	10. 34+16=		20. ? + 7 = 15	

Module 4 – Add and Subtract 10's.

Lesson	Draw and solve	Code	Retrieval Practice – 10 mins per day	Code
1	1. 23+10= 2. 49-10= 3. 50-10= 4. 10+34= 5. 45-?=35 6. 34+10= 7. 10+28= 8. ?+10=36 9. ?+24=34 10. 37+?=47		11. 26+6= 12. 33-4= 13. 43-6= 14. ?-6=26 15. 35+?=41 16. 26+?=31 17. 29-?=25 18. 27 + 10 = 19. 28 + 1 = 20. 39 - 2 = 21. 32 +? = 40 22. 38 - 20 = 23. 6 + 7 = 24. 12 - 6 =	
2	1. 13+10= 2. 39-10= 3. 40-10= 4. 10+24= 5. 41-?=31 6. 32+10= 7. 10+38= 8. ?+10=26 9. ?+24=44 10. 35+?=45		11. 43 - 10 = 12. 29 + 10 = 13. 40 + 10 = 14. 34 - 10 = 15. 25 - ? = 15 16. 29 - 10 = 17. ? + 15 = 37 18. 36 - 10 = 19. 39 - 20 = 20. 22 +? = 42	
3	1. 33+10=		11. 43 - 1 =	

© Copyright 2024 Mathtastic: Tracy Ashbridge. All rights reserved

	2. 39-20=		12. 32 - 2 =	
	3. 40-20=		13. 6 + 37 =	
	4. 10+14=		14. 37 + ? = 40	
	5. 41-?=21		15. 18 + ? = 20	
	6. 12+10=		16. ? + 3 = 27	
	7. 10+16=		17. 5 + 18 =	
	8. ?+10=45		18. ? + 7 = 30	
	9. ?+34=44		19. 15 + ? = 30	
	10. 25+?=45		20. 24 + ? = 50	
4	1. 36+10=		11. 17 - 8 =	
	2. 39-30=		12. 25 + 2 =	
	3. 50-10=		13. 37 - 3 =	
	4. 10+18=		14. 8 + 9 =	
	5. 49-?=29		15. 20 - 10 =	
	6. 32+10=		16. 8 + 6 =	
	7. 10+19=		17. 17 - 10 =	
	8. ?+10=34		18. 19 -? = 10	
	9. ?+24=44		19. 3 + 10 =	
	10. 15+?=45		20. ? + 4 = 12	
5	1. 10+25=		11. 32 - 3 =	
	2. ?+10=14		12. 22 + 2 =	
	3. ?+20=22		13. 42 +? =	
	4. 14+?=34		14. ? + 8 = 40	
	5. 25+10=		15. 29 - 20 =	
	6. 33-30=		16. 40 - 1 =	
	7. 50-30=		17. 23 -? = 18	
	8. 10+38=		18. 18 + 22 =	
	9. 49-?=19		19. 18 + 10 =	
	10. 22+10=		20. 17 + 30 =	

© Copyright 2024 Mathtastic: Tracy Ashbridge. All rights reserved

Module 5 – Doubles and Halves

Lesson	Draw and solve	Code	Retrieval Practice – 10 mins per day	Code
1	1. 12+12= 2. 15+15= 3. 21+21= 4. 25+25= 5. Double 23 = 6. 24-12= 7. Half of 28 = 8. Half of 42 = 9. Half of 48 = 10. 44-?=22		11. 27+2= 12. 36-1= 13. 8+25= 14. ?+7=41 15. ?+10= 46 16. 22+22= 17. 24+?=30 18. 16+?=3- 19. 19+20= 20. 14+14=	
2	1. Double 24 = 2. 10+10= 3. Half of 10 = 4. Double 14 = 5. 16+16= 6. 12 ÷ 2 = 7. 20-10 = 8. 22 x 2 = 9. Half of 30 = 10. 42-? = 21		11. 13+13= 12. 25+25= 13. 28-14= 14. 40-20= 15. Double 16 16. 24+24= 17. 21 x 2 = 18. Half of 48 19. 36 ÷ 2 = 20. Double 18	

© Copyright 2024 Mathtastic: Tracy Ashbridge. All rights reserved

3	1. Double 20 = 2. 16 x 2 = = 3. Half of 30 = 4. 28-14 = 5. Half of 26 = 6. 18 ÷ 2 = 7. 11+11 = 8. Half of 46 = 9. 24 ÷ 2 = 10. 25 x 2 =		11. 23+10= 12. 49-10= 13. 43-6= 14. ?-6=26 15. 35+?=41 16. 26+?=31 17. 29-?=25 18. 25+5= 19. 41+9= 20. 34+6 =	
4	1. 18 x 2 = 2. Half of 40 = 3. 22 ÷2 = 4. Double 19 = 5. Double 25 = 6. 16 ÷2 = 7. Half of 32 = 8. 16 x 2 = 9. Double 11 = 10. Half of 14 =		11. 50-10= 12. 10+34= 13. 24+?=32 14. 12+?=20 15. 46+?=50 16. ?+25=40 17. 41+3= 18. 38-2= 19. 16 X2= 20. Double 20 21. Half of 28 22. 26 ÷ 2 =	

5	1. $26 \div 2 =$		11. $45 - ? = 35$	
	2. Half of 34 =		12. $34 + 10 =$	
	3. Double 24 =		13. $29 + ? = 40$	
	4. Double 16 =		14. $15 + 35 =$	
	5. $32 \div 2 =$		15. $34 + 16 =$	
	6. $25 \times 2 =$		16. $31 + ? = 37$	
	7. Double 15 =		17. $37 - ? = 32$	
	8. Half of 36 =		18. $? + 5 = 44$	
	9. Half of 50 =		19. $? - 4 = 17$	
	10. $18 \div 2 =$		20. $45 - ? = 35$	
			21. $23 \times 2 =$	
			22. $46 \div 2 =$	

© Copyright 2024 Mathtastic: Tracy Ashbridge. All rights reserved

Module 6 – Near Doubles

Lesson	Draw and solve	Code	Retrieval Practice – 10 mins per day	Code
1	1. 24+25 = 2. 15+15 +1 = 3. 21+22 = 4. 22+22 = 1 = 5. 15+15 -1 = 6. 18+19 = 7. 11+13 = 8. 16+18 = 9. 20+22 =, 10. 24+26 =		11. 21 + 3 = 12. 29 + 6 = 13. 24 + 16 = 14. 27 + 20 = 15. 16 x 2 = 16. 19 - 1 = 17. 43 - 5 = 18. 38 + ? = 40 19. 38 - 10 = 20. 46 ÷ 2 = 21. 17 + 18 = 22. 10 + 2 =	
2	1. 21+20 = 2. 18+17 = 3. 14+12 = 4. 11+12 = 5. 18+20 = 6. 16+17 = 7. 13+15 = 8. 23+25 = 9. 18+19 = 10. 19+18 =		11. 23 + 22 = 12. 14 + 14 + 1 = 13. 19 + 20 = 14. 18 + 18 - 1 = 15. 12 + 14 = 16. 19 + 19 - 1 = 17. 14 + 13 = 18. 24 + 26 = 19. 15 + 17 = 20. 17 + 18 =	

© Copyright 2024 Mathtastic: Tracy Ashbridge. All rights reserved

3	1. 12+13 =		11. 17 + 13 =	
	2. ?+24 =47		12. 28 + 20 =	
	3. 14+? =27		13. double 23	
	4. 17+19 =		14. 22 + 23 =	
	5. 19+?= 39		15. 12 + 11 =	
	6. ?+26 = 50		16. 31 + ? = 40	
	7. 13+? = 27		17. 19 + ? = 49	
	8. 21+?= 43		18. Half of 38	
	9. 15+?=31		19. 16 + 17 =	
	10. 24+24+1=		20. 26 ÷ 2 =	
4	1. 24+25 =		11. 42 + 3 =	
	2. 15+15 +1 =		12. 27 + 6 =	
	3. 21+22 =		13. 13 + 14 =	
	4. 22+?= 43		14. 10 + 7 =	
	5. 15+?=31		15. 8 + 4 =	
	6. 24+24+1=		16. 28 - 1 =	
	7. 16+17 =		17. 38 - 5 =	
	8. 13+15 =		18. 23 + 22 =	
	9. 23+25 =		19. 19 - 10 =	
	10. 18+19 =		20. ? + 5 = 12	
5	1. 19+18 =		11. ? + 7 = 40	
	2. 17+19 =		12. 22 + ? = 30	
	3. 19+?= 39		13. 33 + 10 =	
	4. 18+19 =		14. ? + 20 = 44	
	5. 11+13 =		15. Double 18	
	6. 16+18 =		16. Half of 34	
	7. 21+22 =		17. 23 + 24 =	
	8. 17+18 =		18. 17 + ? = 35	
	9. 20+22 =		19. 48 + ? = 50	
	10. 24+26 =		20. ? + 6 = 36	

Module 7 – partition by place value

Lesson	Draw and solve	Code	Retrieval Practice – 10 mins per day	Code
1	1. 10+3= 2. 40+4= 3. 3+30= 4. 20+?=25 5. ?+30=34 6. 20+?=26 7. 7+?=17 8. 9+30= 9. 8+40= 10. ?+3=43		11. 25 + 3 = 12. 47 - 1 = 13. 32 + 18 = 14. 42 - 20 = 15. 25 + 24 = 16. 6 + 26 = 17. 42 + ? = 50 18. 47 - 30 = 19. 22 x 2 = 20. 32 ÷ 2 =	
2	1. 40+3= 2. 4+20= 3. 3+20= 4. 20+?=27 5. ?+30=30 6. 20+?=21 7. 7+?=47 8. 9+10= 9. 1+40= 10. ?+4=14		11. 20 + 6 = 12. 30 + 4 = 13. ? + 20 = 44 14. 7 + ? = 27 15. 1 + 40 = 16. 9 + 20 = 17. 40 + ? = 43 18. 30 + ? = 36 19. 5 + 30 = 20. ? + 5 = 35	

© Copyright 2024 Mathtastic: Tracy Ashbridge. All rights reserved

3	1. 14+10= 2. 33+10- 3. 10+29= 4. 20+21= 5. 30+?=40 6. 40+?=45 7. 24+20= 8. 30+5= 9. ?+25=35 10. ?+36=46		11. 36 + 10 = 12. 45 - 20 = 13. Double 21 14. Half of 50 15. 16 + 18 = 16. 37 - 30 = 17. 15 x 2 = 18. 24 ÷ 2 = 19. 29 + 6 = 20. 37 + ? = 42	
4	1. 17+10= 2. 26+20- 3. 10+39= 4. 10+21= 5. 30+?=50 6. 40+?=50 7. 14+20= 8. 30+8= 9. ?+15=35 10. ?+16=46		11. 18 + 3 = 12. 32 - 4 = 13. 43 + ? = 50 14. 27 + 21 = 15. 29 + 6 = 16. 29 - 3 = 17. 47 - 6 = 18. ? + 30 = 33 19. 16 + 13 = 20. 35 - 6 =	
5	1. 20+?=35 2. 24+30= 3. 30+?=40 4. ?+25=25 5. ?+16=36 6. 14+30= 7. 12+10- 8. 10+19= 9. 30+20= 10. 20+?=40		11. 24+16= 12. 36+10= 13. 45-20= 14. Double 16 15. 25 X 2 = 16. Half of 50 17. 20 ÷ 2= 18. 17+19= 19. 25+24= 20. 32+12=	

© Copyright 2024 Mathtastic: Tracy Ashbridge. All rights reserved

Module 8 – Add and Subtract by Compensation.

Lesson	Draw and solve	Code	Retrieval Practice – 10 mins per day	Code
1	1. 16+5= 2. 27+?=32 3. 28+4= 4. 29+4 5. 36+5= 6. 38+6= 7. 29+5= 8. 26+?=31 9. 17+4= 10. 40+10=		11. 25+3= 12. 19-1= 13. 41-3= 14. 46-7= 15. Double 16 16. 27-2= 17. 43+2= 18. 28+6= 19. 33-5= 20. 23 x 2 = 21. 31+12= 22. 24+26= 23. 18+19= 24. 4+26= 25. ?+38=40 26. 24+20= 27. 17+30= 28. 46 ÷2= 29. 36÷2=	
2	1. 18+6= 2. 27+4= 3. 28+5= 4. 5+19= 5. 29+?=34 6. 26+5= 7. 37+?=42 8. 18+?=22 9. 38+?=42 10. 35+?=41		11. 29+5= 12. 38+5= 13. 27+5= 14. 18+6= 15. 35+?=42 16. 37+?=43 17. 19+6= 18. 38+4= 19. 39+?=43 20. 16+?=22	

3	1. 16+6=		11. Double 11	
	2. 6+19=		12. 22 x 2=	
	3. 28+6=		13. Half of 46	
	4. 38+?=44		14. 38 ÷ 2=	
	5. ?+38=41		15. 36+5=	
	6. 38+?=44		16. 13+14=	
	7. 36+6=		17. 21+22=	
	8. 37+4=		18. 17+22=	
	9. 29+6=		19. 33+14=	
	10. 39+?=43		20. 18+4=	
4	1. 17+?=22		11. 41+3=	
	2. 18+?=23		12. 50-6=	
	3. 37+6=		13. 36+?=40	
	4. 38+5=		14. 44-21=	
	5. 39+?=45		15. 20+25=	
	6. 39+?=41		16. 39-2=	
	7. ?+36=42		17. 32+5=	
	8. 38+?=43		18. 18+22=	
	9. 37+5=		19. 12+30=	
	10. ?+28=32		20. 37+5=	
5	1. 17+6=		11. 16 x 2=	
	2. 27+?=33		12. Half of 38=	
	3. 26+6=		13. Double 23	
	4. 38+4=		14. 18 ÷ 2=	
	5. 39+?=44		15. 14+15=	
	6. ?+36=41		16. 22+26 =	
	7. 38+?=44		17. 31+17=	
	8. 29+?=34		18. 18 +24=	
	9. ?+26=31		19. 27+16=	
	10. 36+?=42		20. 16+16-1=	

© Copyright 2024 Mathtastic: Tracy Ashbridge. All rights reserved

Level 3
Numbers to 50
Problem Solving book

How to use these cards

How to use these problem cards

Each card is written to meet each of the different ways of presenting a problem. These are written at the bottom of each page for teacher reference.

The problems are deliberately written with each piece of the problem written on a separate line. This way you can cover up the rest of the problem and reveal one line at a time.

Bet lines - this is a technique to help students to think about the problem. You show one line at a time and as students to "bet" what the problem is going to ask them to do. They review their ideas as more information is revealed.

For each problem there are 2 extra sets of numbers which can be used instead for extra practice of that problem type - at the bottom of the page in brackets next to the problem type)

Recording Page

Use this page to record how the student managed with each different problem type. Which can they do easily, and which need more practice?

Join – result unknown 6+2=?	Join – change unknown 6+?=8	Join – start unknown ?+2=8
Separate – result unknown 9-5=?	Separate – change unknown 9-?=4	Separate – start unknown ?-5=4
Part-part-whole – whole unknown 5+4=?	Part-part-whole – part unknown ?+4=9	
Compare – difference unknown 7-2=?	Compare – compared set unknown 7-?=5	Compare – referent unknown ?-5=2

© Copyright 2024 Mathtastic: Tracy Ashbridge. All rights reserved

Module 1 - Add and Subtract 0,1,2,3

Georgie had 25 crayons.

Sam gave her 2 more for her birthday.

How many crayons does she have now?

Join - result unknown (31, 43, 29)
Mathtastic Level 3 - Numbers to 50
add and subtract 0,1,2,3
www.tracyashbridge.com

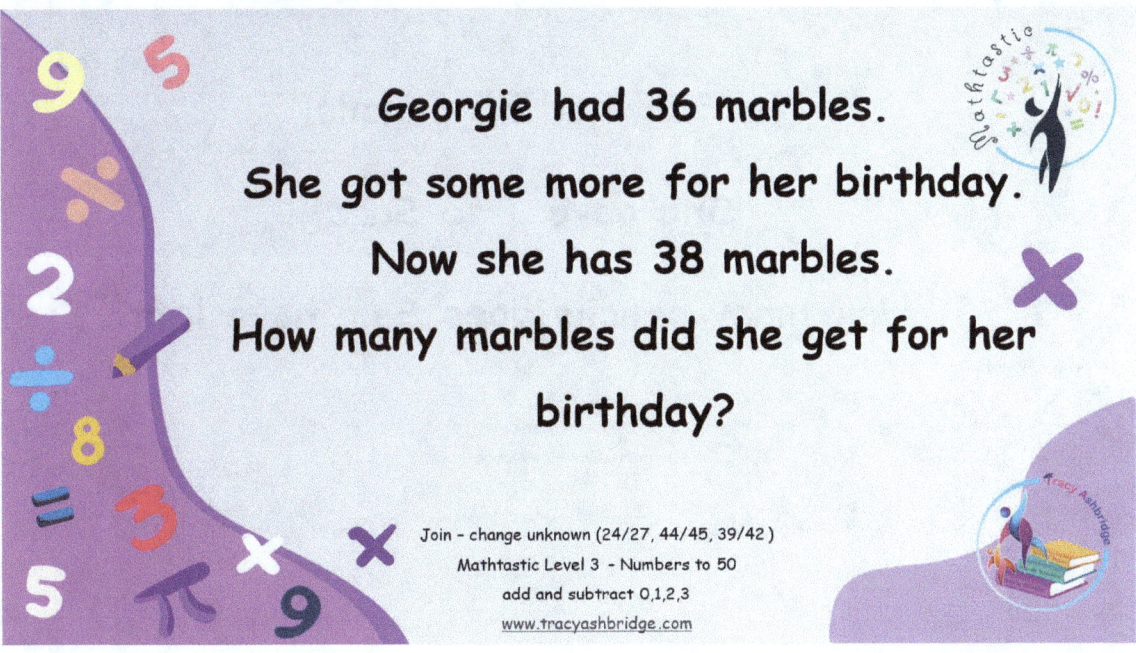

Georgie had 36 marbles.
She got some more for her birthday.
Now she has 38 marbles.
How many marbles did she get for her birthday?

Join - change unknown (24/27, 44/45, 39/42)
Mathtastic Level 3 - Numbers to 50
add and subtract 0,1,2,3
www.tracyashbridge.com

Georgie had some crayons.
Her friend gave her 1 more.
Then she had 45 crayons.
How many crayons did she have at the beginning?

Join – start unknown (3/32, 2/28, 1/30)
Mathtastic Level 3 – Numbers to 50
add and subtract 0,1,2,3
www.tracyashbridge.com

Sam had 35 pencils.
She gave 1 to Sarah.
How many pencils does Sam have left?

Separate – result unknown (27/2, 41/3, 31/2)
Mathtastic Level 3 – Numbers to 50
add and subtract 0,1,2,3
www.tracyashbridge.com

© Copyright 2024 Mathtastic: Tracy Ashbridge. All rights reserved

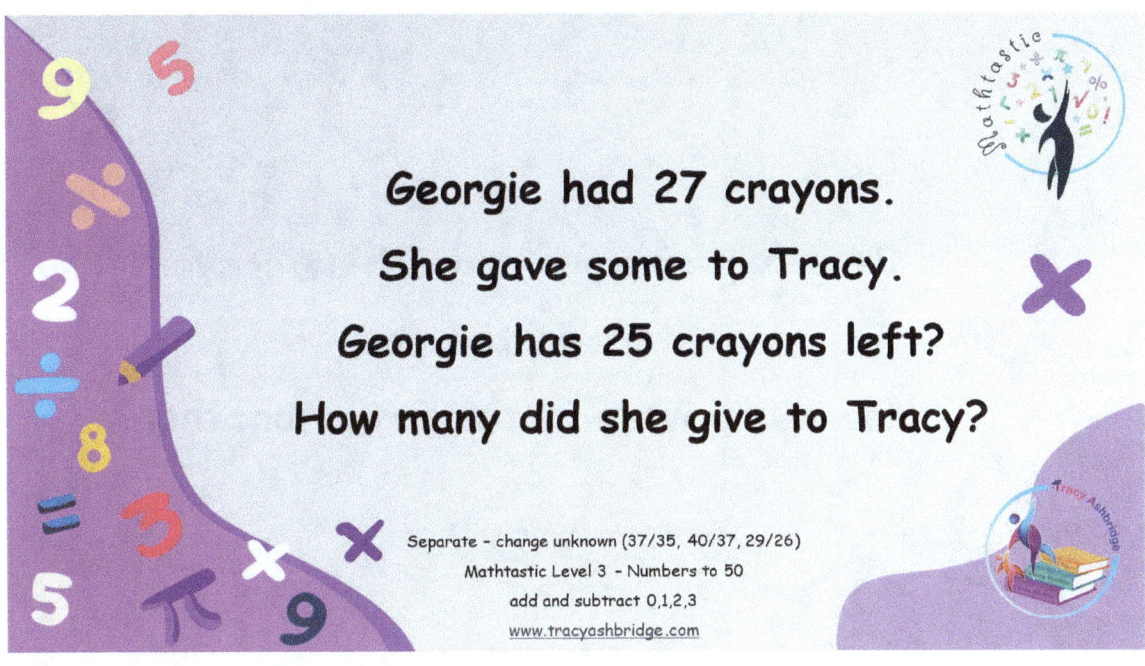

Georgie had 27 crayons.
She gave some to Tracy.
Georgie has 25 crayons left?
How many did she give to Tracy?

Separate – change unknown (37/35, 40/37, 29/26)
Mathtastic Level 3 – Numbers to 50
add and subtract 0,1,2,3
www.tracyashbridge.com

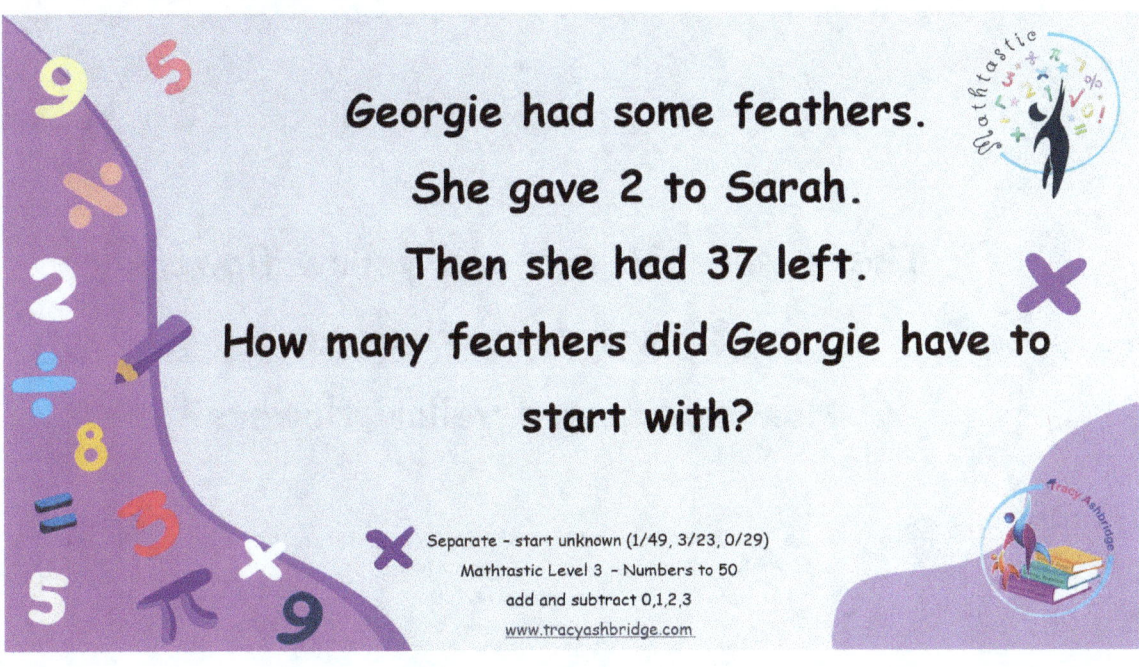

Georgie had some feathers.
She gave 2 to Sarah.
Then she had 37 left.
How many feathers did Georgie have to start with?

Separate – start unknown (1/49, 3/23, 0/29)
Mathtastic Level 3 – Numbers to 50
add and subtract 0,1,2,3
www.tracyashbridge.com

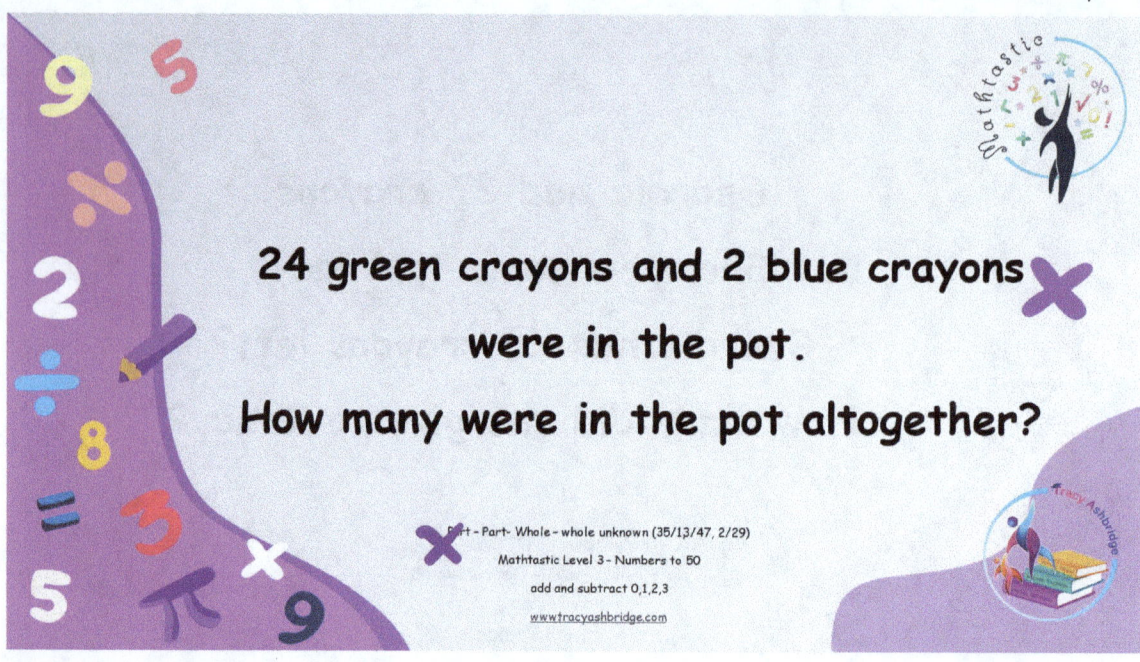

24 green crayons and 2 blue crayons were in the pot.
How many were in the pot altogether?

Part – Part- Whole – whole unknown (35/13/47, 2/29)
Mathtastic Level 3 – Numbers to 50
add and subtract 0,1,2,3
www.tracyashbridge.com

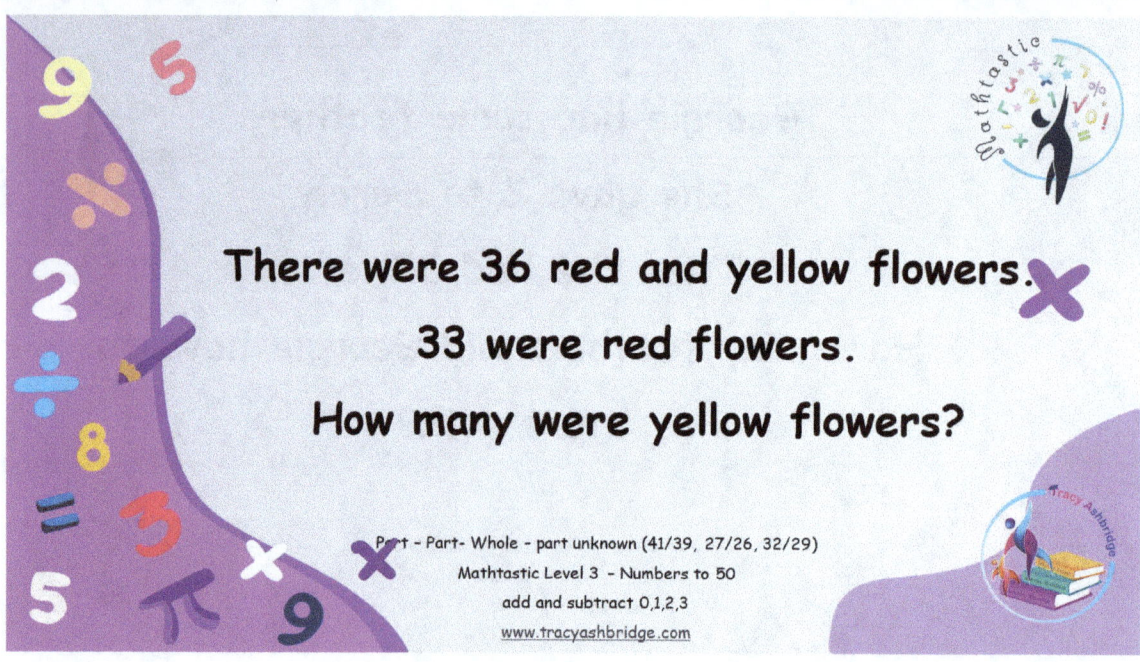

There were 36 red and yellow flowers.
33 were red flowers.
How many were yellow flowers?

Part – Part- Whole – part unknown (41/39, 27/26, 32/29)
Mathtastic Level 3 – Numbers to 50
add and subtract 0,1,2,3
www.tracyashbridge.com

© Copyright 2024 Mathtastic: Tracy Ashbridge. All rights reserved

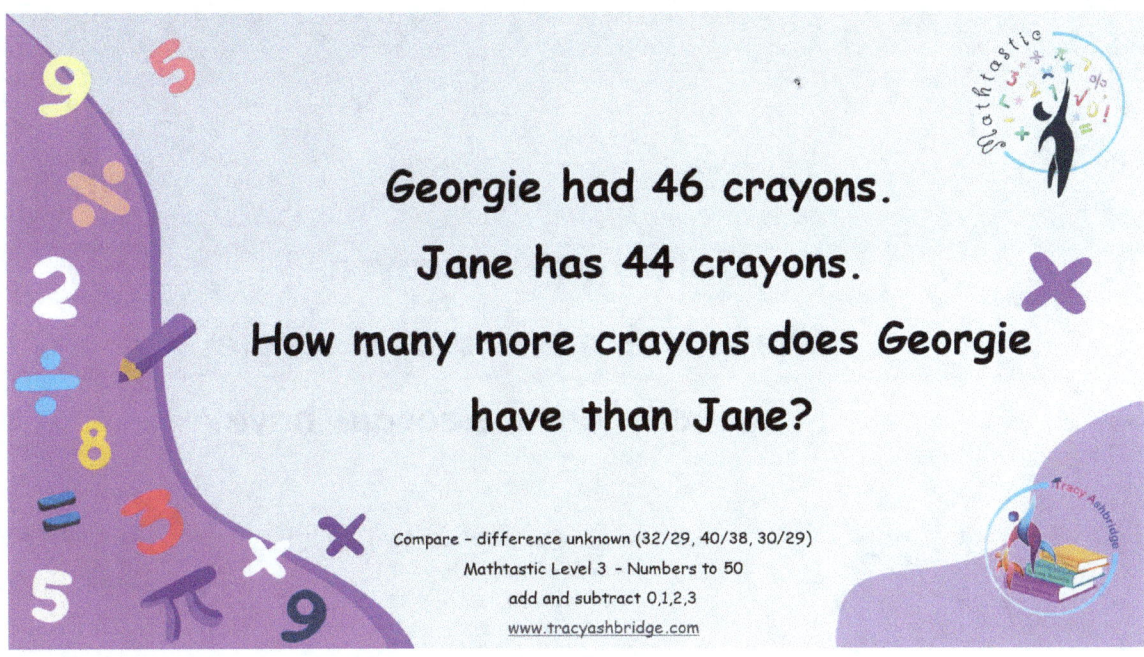

Georgie had 46 crayons.
Jane has 44 crayons.
How many more crayons does Georgie have than Jane?

Compare – difference unknown (32/29, 40/38, 30/29)
Mathtastic Level 3 – Numbers to 50
add and subtract 0,1,2,3
www.tracyashbridge.com

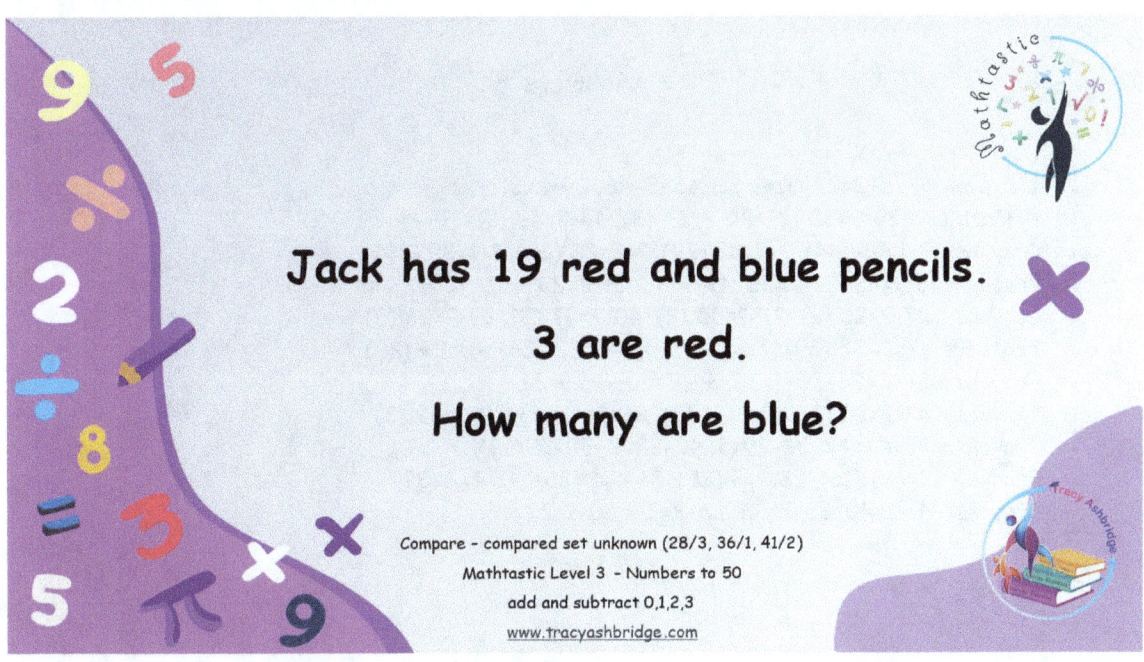

Jack has 19 red and blue pencils.
3 are red.
How many are blue?

Compare – compared set unknown (28/3, 36/1, 41/2)
Mathtastic Level 3 – Numbers to 50
add and subtract 0,1,2,3
www.tracyashbridge.com

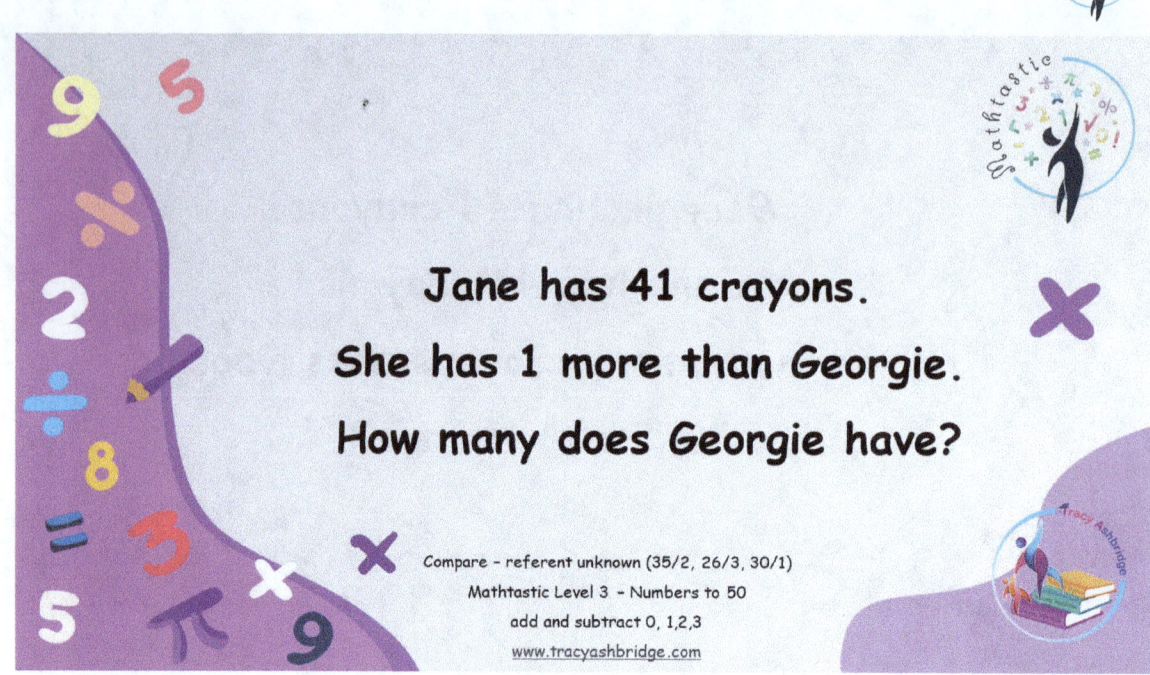

Jane has 41 crayons.
She has 1 more than Georgie.
How many does Georgie have?

Compare – referent unknown (35/2, 26/3, 30/1)
Mathtastic Level 3 – Numbers to 50
add and subtract 0, 1,2,3
www.tracyashbridge.com

Answers

- 1. Crayons – 25+2=27, 31+1=32, 43+3 -46, 29+2=31
- 2. Marbles – 36+?=38 (2), 24+?=27 (3), 44+?=45 (1), 39+?=42 (3)
- Crayons – ?+1=45 (44), ?+3=32 (29), ?+2=28 (26), ?+1=30 (29)
- Pencils – 35-1=43, 27-2=25, 41-3=38, 31-2=29
- Crayons – 27-?=25 (2), 37-?=35 (2), 40-?=37 (3), 29-?=26 (3)
- Feathers – ?=2=37 (39), ?-1=49 (50), ?-3=23 (26), -0=29 (29)
- Crayons – 24+2=26, 35+1=36, 3+47=50, 2+29=31
- Flowers – 33+?=36 (3), 39+?=41 (2), 26+?=41 (2), 29+?=32 (3)
- Crayons – 46-44=42, 32-29=3, 40-38=2, 30-29=1
- Pencils – 19-?=3 (16), 28-?=3 (25), 36-?=1 (35), 41-?=2 (39)
- Crayons – 41-1=40, 35-2=33, 26-3=23, 30-1=29

Module 2 - Add from largest, subtract by counting back

Jake had 25 Pokémon cards. His neighbor gave him 6 more as a gift. How many Pokémon cards does he have now?

Join – result unknown (37/5, 46/4, 29/5)
Mathtastic Level 3 – Numbers to 50
Add from largest, subtract by counting back
www.tracyashbridge.com

Oliver had 20 toy trains. He received some more as a birthday gift. Now, he has 27 toy trains. How many toy trains did he receive?

Join – change unknown (35/9, 41/7, 27/6)
Mathtastic Level 3 – Numbers to 50
Add from largest, subtract by counting back
www.tracyashbridge.com

Emily had some puzzle pieces.
Her friend gave her 5 more.
Now she has 28 puzzle pieces.
How many puzzle pieces did she have before her friend gave her more?

Join - start unknown (3/37, 8/41, 7/36)
Mathtastic Level 3 - Numbers to 50
Add from largest, subtract by counting back
www.tracyashbridge.com

Tom had 45 stamps.
He gave 6 stamps to his sister.
How many stamps does Tom have now?

Separate - result unknown (38/6, 27/8, 32/7)
Mathtastic Level 3 - Numbers to 50
Add from largest, subtract by counting back
www.tracyashbridge.com

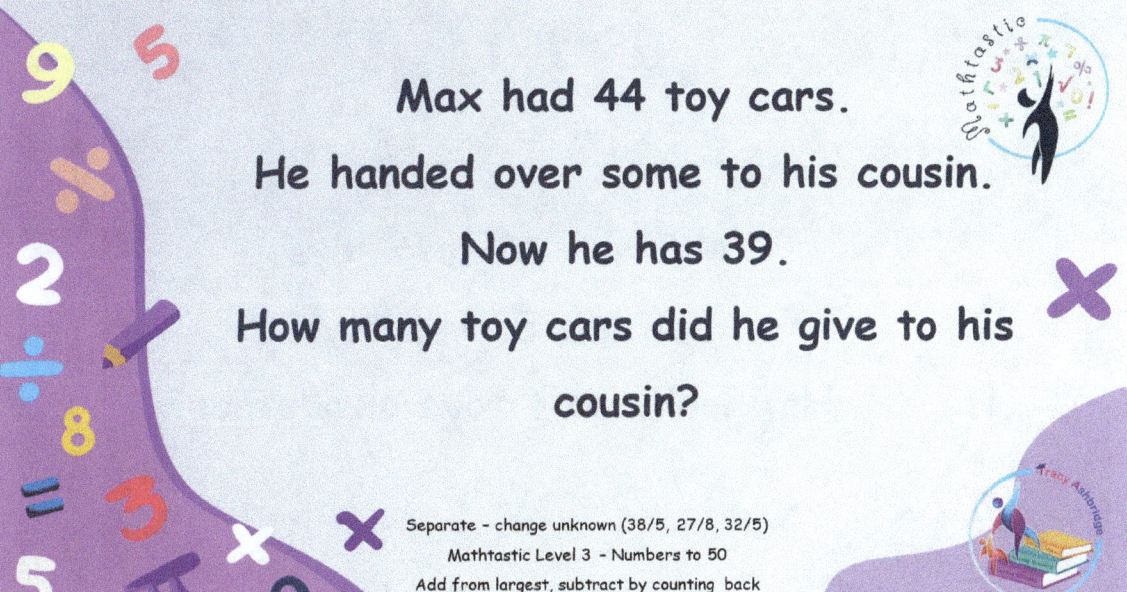

Max had 44 toy cars.
He handed over some to his cousin.
Now he has 39.
How many toy cars did he give to his cousin?

Separate – change unknown (38/5, 27/8, 32/5)
Mathtastic Level 3 – Numbers to 50
Add from largest, subtract by counting back
www.tracyashbridge.com

Lily had some books.
She lent 2 to her neighbour.
Now Lily has 47 books.
How many books did her Lily have at the start?

Separate – start unknown (26/4, 6/39, 29/3)
Mathtastic Level 3 – Numbers to 50
Add from largest, subtract by counting back
www.tracyashbridge.com

© Copyright 2024 Mathtastic: Tracy Ashbridge. All rights reserved

Mia has 28 soft toys.
Ben has 5 soft toys.
How many soft toys altogether?

Part – Part- Whole – whole unknown (24/5, 45/4, 6/29)
Mathtastic Level 3 – Numbers to 50
Add from largest, subtract by counting back
www.tracyashbridge.com

Emma had 36 toy trains.
4 broke down.
How many toy trains are left?

Part – Part- Whole – part unknown (49/8, 24/5, 33/6)
Mathtastic Level 3 – Numbers to 50
Add from largest, subtract by counting back
www.tracyashbridge.com

Jake had 25 cookies.
He gave 7 to a friend.
How many cookies are left?

Compare – difference unknown (47/8, 35/6, 43/5)
Mathtastic Level 3 – Numbers to 50
Add from largest, subtract by counting back
www.tracyashbridge.com

Max had 42 ice cream cones.
He invited 34 friends for an ice cream party.
How many extra ice cream cones does he have?

Compare – compared set unknown (29/34, 41/48, 26/19)
Mathtastic Level 3 – Numbers to 50
Add from largest, subtract by counting back
www.tracyashbridge.com

© Copyright 2024 Mathtastic: Tracy Ashbridge. All rights reserved

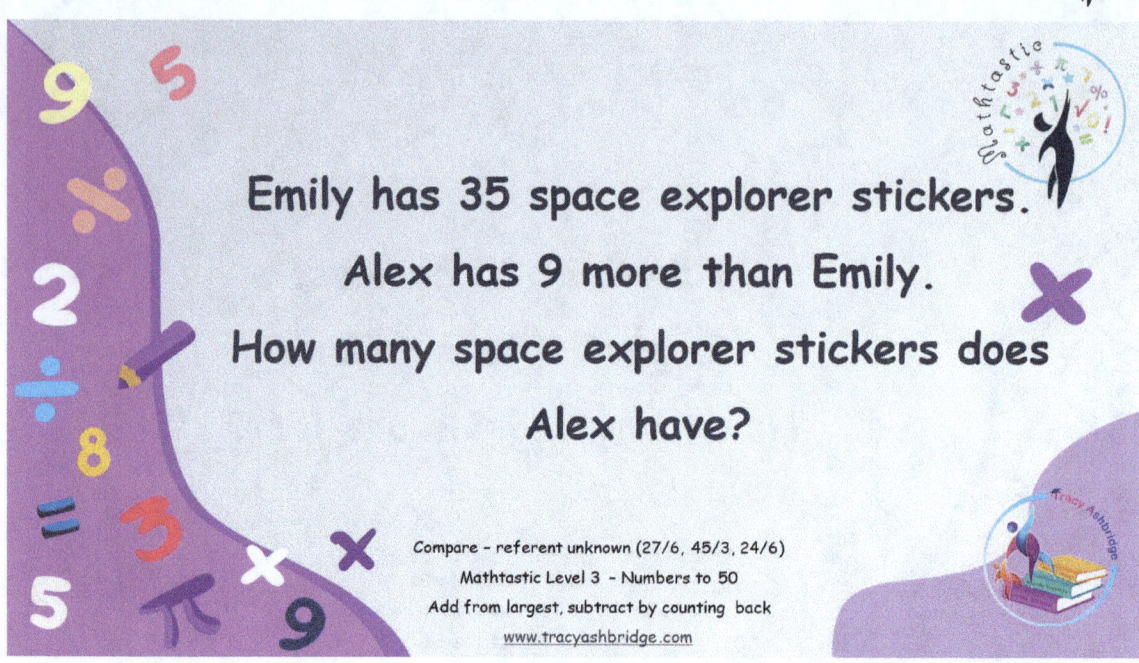

Emily has 35 space explorer stickers. Alex has 9 more than Emily. How many space explorer stickers does Alex have?

Compare – referent unknown (27/6, 45/3, 24/6)
Mathtastic Level 3 – Numbers to 50
Add from largest, subtract by counting back
www.tracyashbridge.com

Answers

- Pokémon – 25+6=31, 37+5=42, 46+4=48, 29+5=36
- Trains - 20+?=27 (7), 5+?=44 (9), 41+?=48 (7), 27+?=33 (6)
- Puzzle - ?+3=28 (35), ?+3=37 (34), ?+8+41 (33), ?+7=36 (29)
- Stamps – 45-6=39, 38-6-32, 27-8=19, 32-7=25
- Cars – 44-?=39 (5), 27-?=8 (19), 38-?=5 (33), 32-?=5 (27)
- Books - ?-2=47 (49), ?-4=39 (43), ?=26=4 (30), ?=29=3 (32)
- Toys – 28+5=33, 24+5=29, 45+4=49, 6+29=35
- 36-4=32, 49-8=41, 24-5=19, 33-6=27
- Cookies – 25-7=18, 47-8=39, 35-6=29, 43-5=38
- Ice cream cones – 42-?=34 (8), 34-?=29 (5), 48-?=41 (7), 26-?=19 (7)
- Stickers – 35+9=44, 27+6=33, 45+3=48, 24+6=30

Module 3 - Rainbow Facts

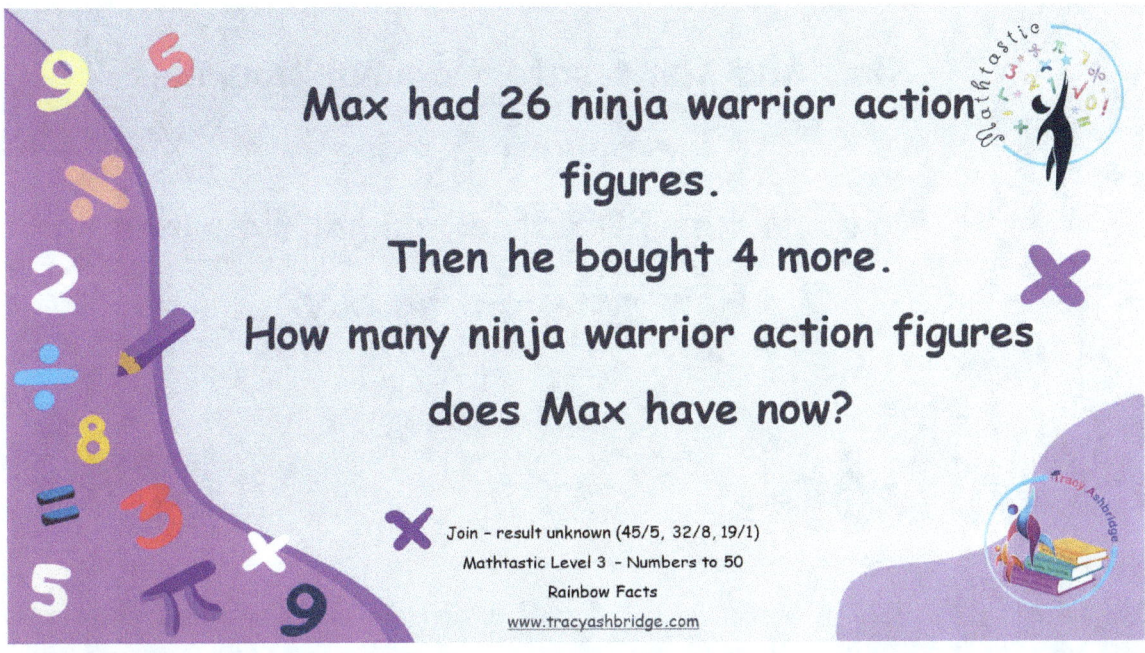

Max had 26 ninja warrior action figures.
Then he bought 4 more.
How many ninja warrior action figures does Max have now?

Join – result unknown (45/5, 32/8, 19/1)
Mathtastic Level 3 – Numbers to 50
Rainbow Facts
www.tracyashbridge.com

Jake had 20 magical creature cards.
He gave some to his neighbour.
Jake now has 14 magical creature cards.
How many did Jake give his neighbour?

Join – change unknown (40/7, 30/3, 50/45)
Mathtastic Level 3 – Numbers to 50
Rainbow Facts
www.tracyashbridge.com

Alex had some safari animal figurines.
He bought 5 more.
Now, he has 30 safari animal figurines.
How many did he buy?

Join – start unknown (3/40, 6/20, 8/50)
Mathtastic Level 3 – Numbers to 50
Rainbow Facts
www.tracyashbridge.com

Alex had 40 robot parts.
He sold 7.
How many robot parts does Alex have now?

Separate – result unknown (30/4, 20/1, 50/6)
Mathtastic Level 3 – Numbers to 50
Rainbow Facts
www.tracyashbridge.com

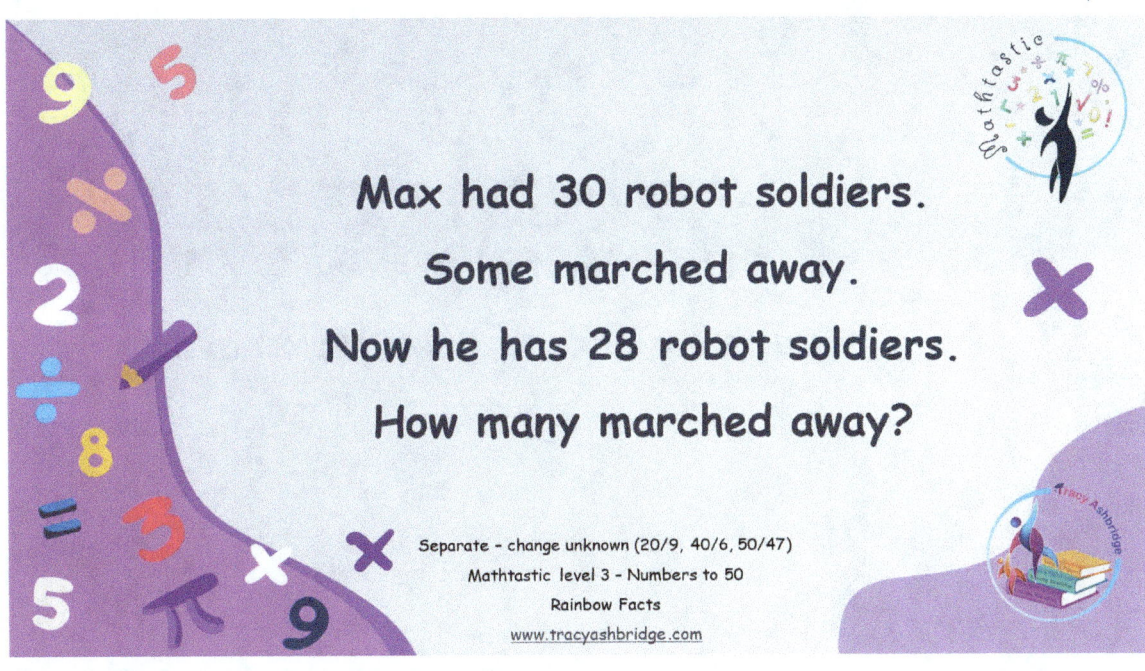

Max had 30 robot soldiers.
Some marched away.
Now he has 28 robot soldiers.
How many marched away?

Separate – change unknown (20/9, 40/6, 50/47)
Mathtastic level 3 – Numbers to 50
Rainbow Facts
www.tracyashbridge.com

Alex had some superhero gadgets.
His friend borrowed 1.
Now he only has 29.
How many did he have at the start?

Separate – start unknown (5/45, 32/8, 6/20)
Mathtastic Level 3 – Numbers to 50
Rainbow Facts
www.tracyashbridge.com

© Copyright 2024 Mathtastic: Tracy Ashbridge. All rights reserved

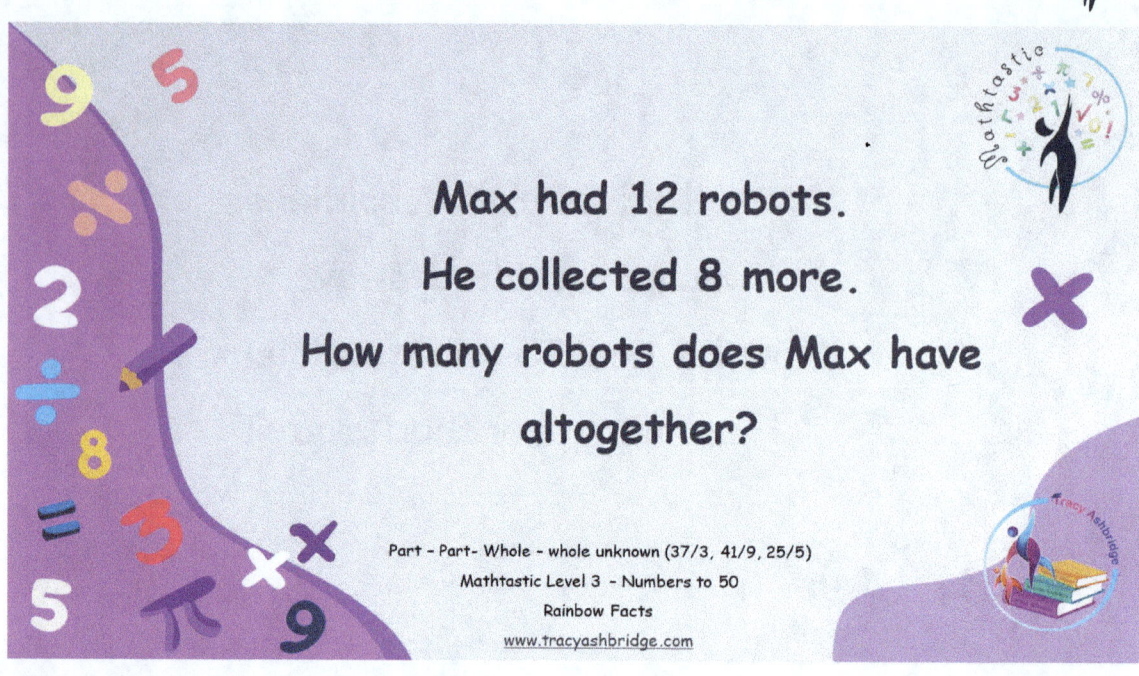

Max had 12 robots.
He collected 8 more.
How many robots does Max have altogether?

Part – Part- Whole – whole unknown (37/3, 41/9, 25/5)
Mathtastic Level 3 – Numbers to 50
Rainbow Facts
www.tracyashbridge.com

Jake hid 50 pirate coins.
His sister found 42.
How many coins were not found?

Part – Part- Whole – part unknown (30/21, 40/36, 20/14)
Mathtastic Level 3 – Numbers to 50
Rainbow Facts
www.tracyashbridge.com

© Copyright 2024 Mathtastic: Tracy Ashbridge. All rights reserved

Ava has 30 fossils.
Some are incomplete.
21 are whole.
How many fossils are incomplete?

Compare – difference unknown (40/36, 20/15, 50/43)
Mathtastic Level 3 – Numbers to 50
Rainbow Facts
www.tracyashbridge.com

Olivia has 20 pirates.
Some have swords.
15 do not have swords.
How many pirates do not have swords?

Compare – compared set unknown (40/32, 30/28, 50/49)
Mathtastic Level 3 – Numbers to 50
Rainbow Facts
www.tracyashbridge.com

© Copyright 2024 Mathtastic: Tracy Ashbridge. All rights reserved

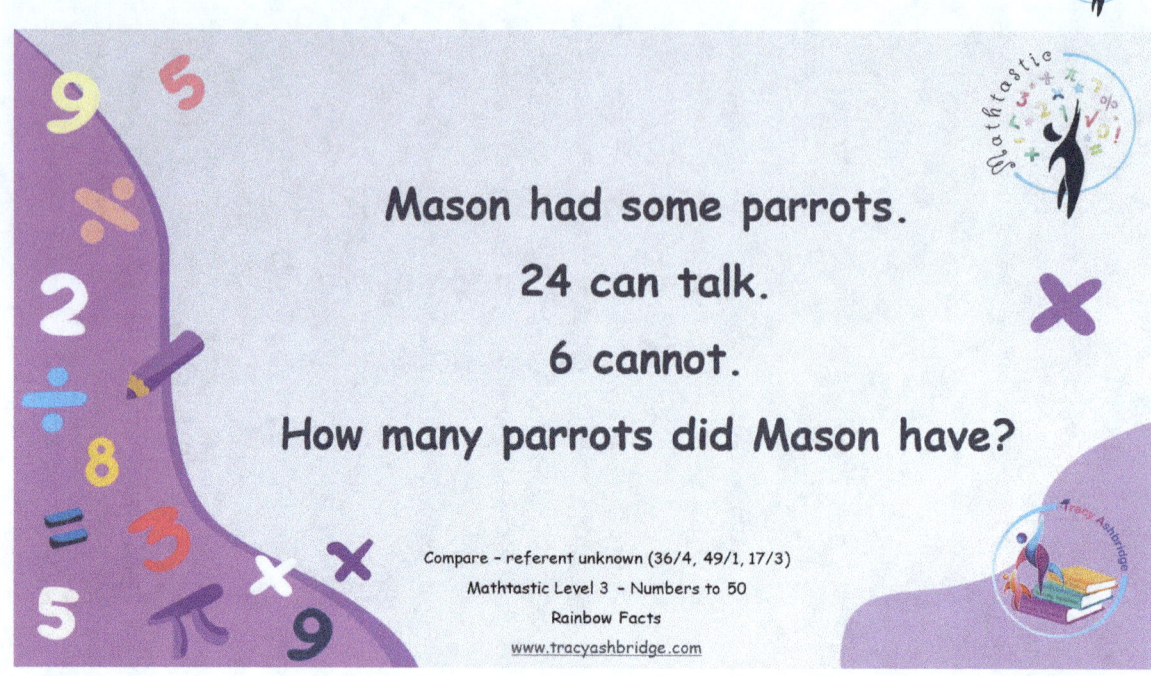

Mason had some parrots.

24 can talk.

6 cannot.

How many parrots did Mason have?

Compare – referent unknown (36/4, 49/1, 17/3)
Mathtastic Level 3 – Numbers to 50
Rainbow Facts
www.tracyashbridge.com

Answers

- Ninja warriors – 26+4=30, 45+5=50, 32+8=40, 19+1=20
- Magical creature cards – 20-?=16 (4), 40-?=7 (33), 30-?=27 (3), 50-?=45 (5)
- Safari animals - ?+5=30 (25), ?+3=40 (37), ?+6=20 (14), ?+8=50 (42)
- Robot parts – 40-7=33, 30-4=26, 20-1=19, 50-6=44
- Robot soldiers – 30-?=28 (2), 20-?=9 (11), 40-?=6 (34), 50-?=47 (3)
- Superhero gadgets - ?-1=29 (30), ?-5=45 (50), ?-32=8 (40), ?-6=20 (26)
- Robots – 12+8=20, 37+3=40, 41+9=50, 25+5=30
- Pirate coins – 50-?=42 (8), 30-?=21 (9), 40-?=36 (4), 20-?=14 (6)
- Fossils – 30-?=21 (9), 40-?=36 (4), 20=?=15 (5), 50 -?=43 (7)
- Pirates – 20-?=15 (5), 40-?=32 (8), 30-?=28 (2), 50-?=49 (1)
- Parrots – 24+6=30, 36+4=40, 49+1=50, 17+3=20

© Copyright 2024 Mathtastic: Tracy Ashbridge. All rights reserved

Module 4 - Add and Subtract 10

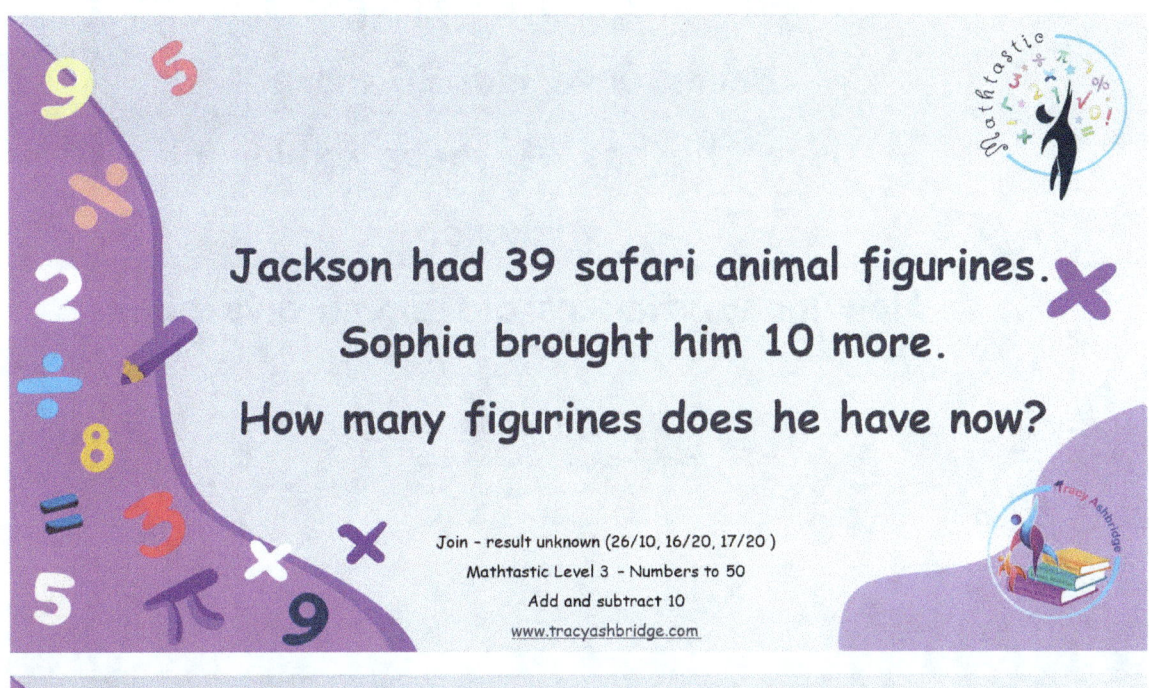

Jackson had 39 safari animal figurines.
Sophia brought him 10 more.
How many figurines does he have now?

Join - result unknown (26/10, 16/20, 17/20)
Mathtastic Level 3 - Numbers to 50
Add and subtract 10
www.tracyashbridge.com

Liam had 32 dinosaur eggs.
Aria found some more.
Now, he has 42 dinosaur eggs.
How many dinosaur eggs did Aria find?

Join - change unknown (10/25, 26/46, 18/28)
Mathtastic Level 3 - Numbers to 50
Add and subtract 10
www.tracyashbridge.com

Liam had some space explorer stickers.
Emma gave him 20 more.
Now, he has 48 space explorer stickers.
How many stickers did Emma give him?

Join – start unknown (10/36, 20/40, 10/45)
Mathtastic Level 3 – Numbers to 50
Add and subtract 10
www.tracyashbridge.com

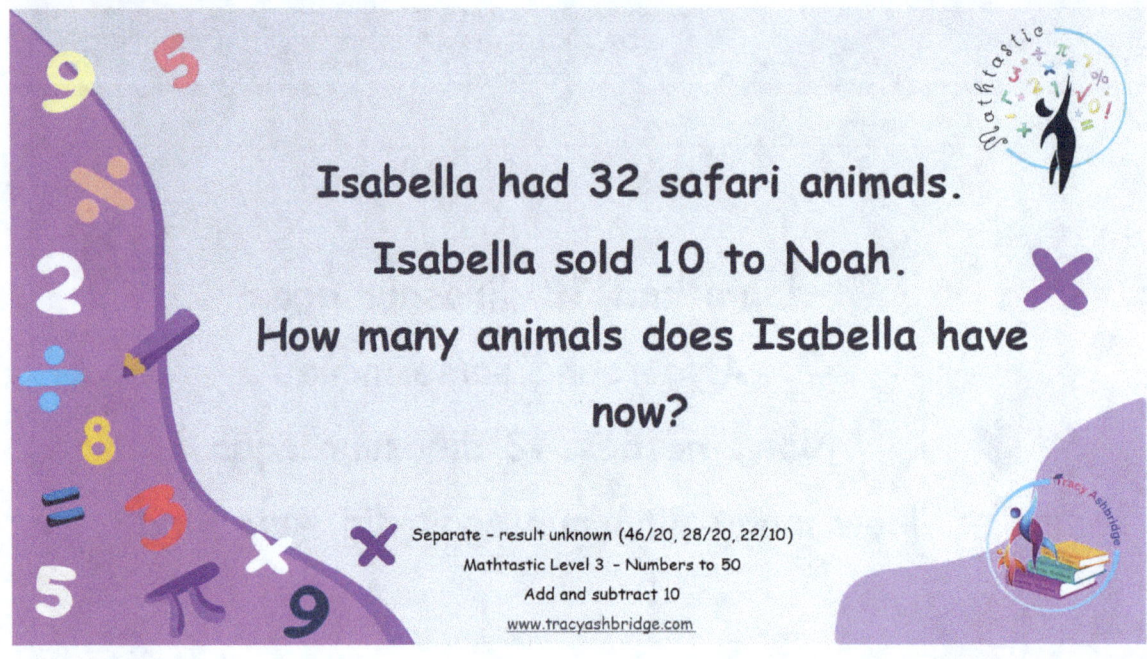

Isabella had 32 safari animals.
Isabella sold 10 to Noah.
How many animals does Isabella have now?

Separate – result unknown (46/20, 28/20, 22/10)
Mathtastic Level 3 – Numbers to 50
Add and subtract 10
www.tracyashbridge.com

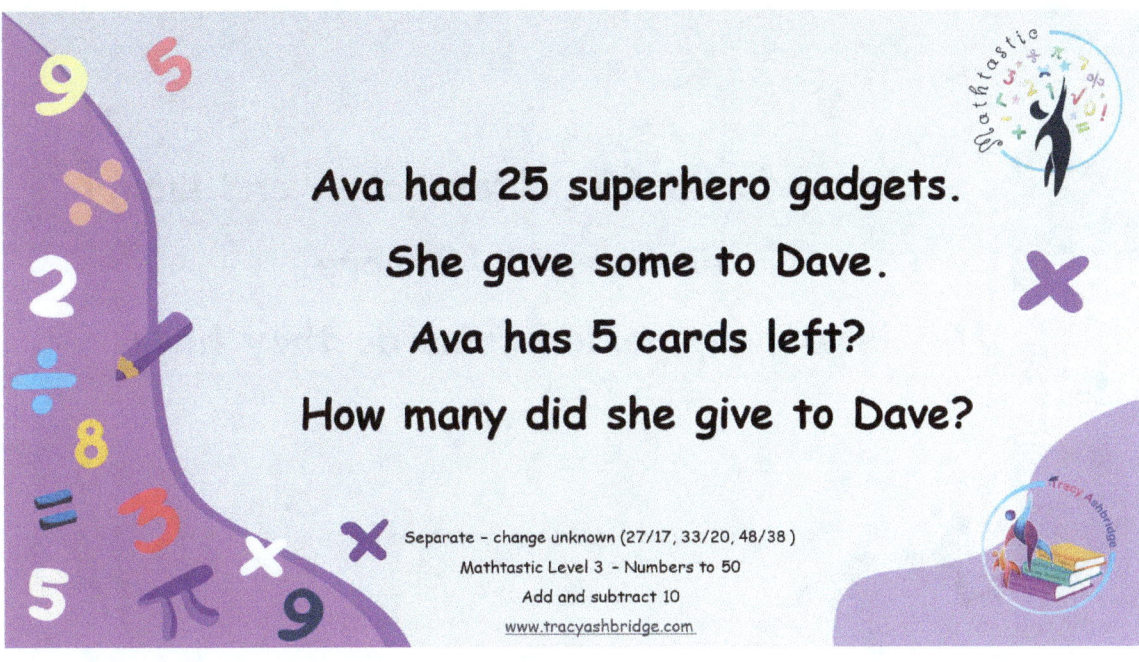

Ava had 25 superhero gadgets.
She gave some to Dave.
Ava has 5 cards left?
How many did she give to Dave?

Separate – change unknown (27/17, 33/20, 48/38)
Mathtastic Level 3 – Numbers to 50
Add and subtract 10
www.tracyashbridge.com

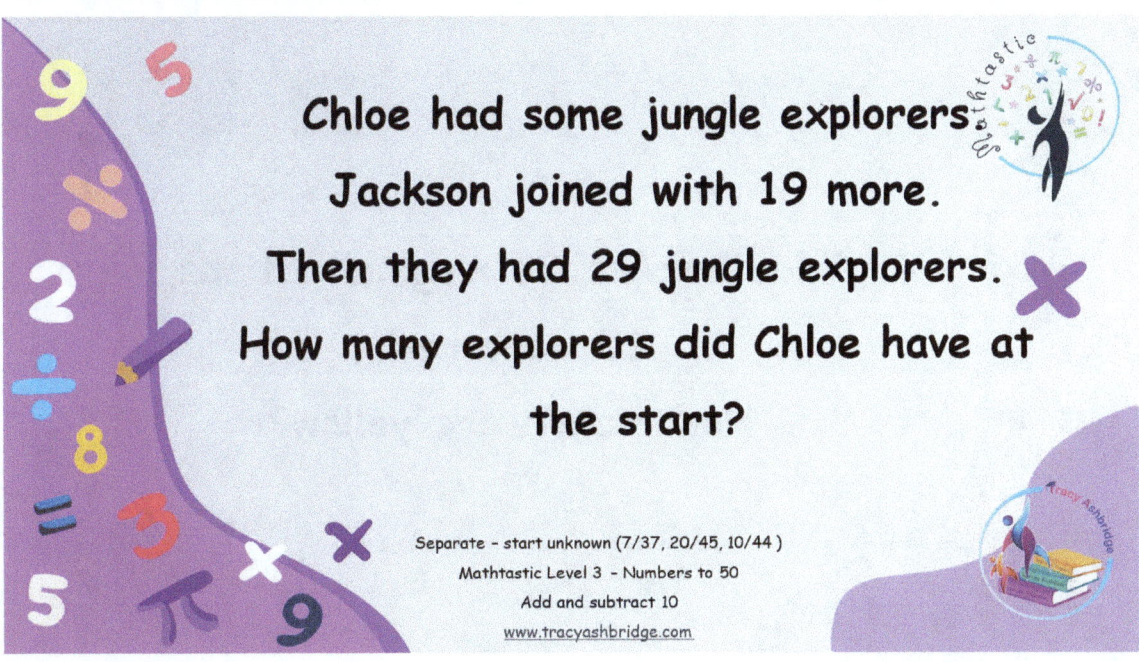

Chloe had some jungle explorers.
Jackson joined with 19 more.
Then they had 29 jungle explorers.
How many explorers did Chloe have at the start?

Separate – start unknown (7/37, 20/45, 10/44)
Mathtastic Level 3 – Numbers to 50
Add and subtract 10
www.tracyashbridge.com

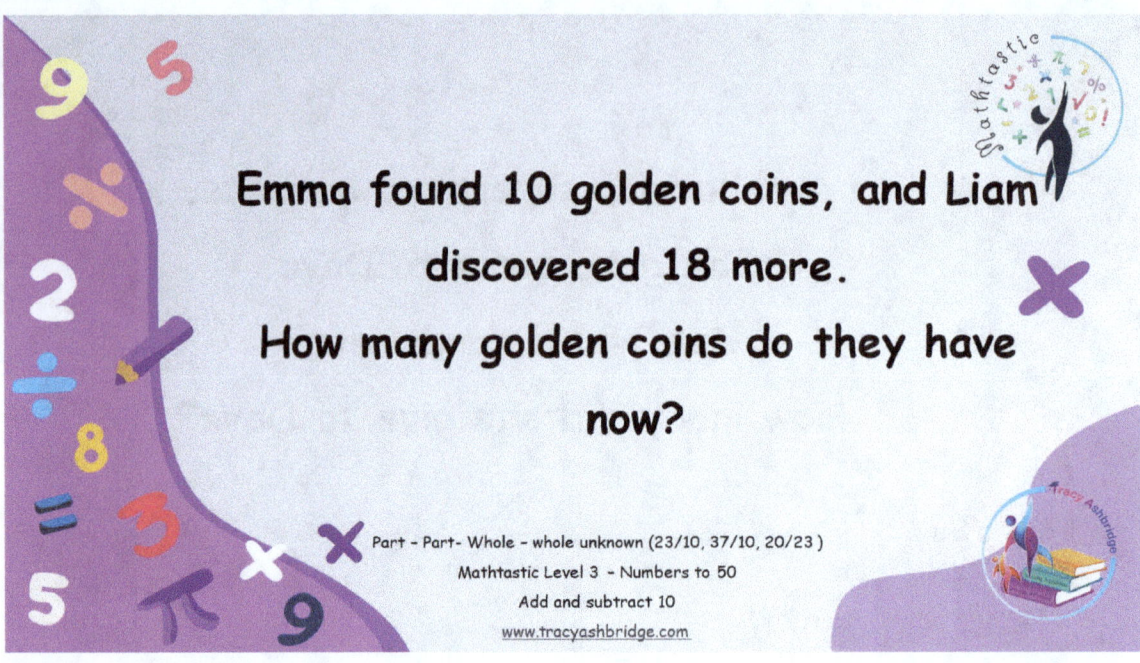

Emma found 10 golden coins, and Liam discovered 18 more.
How many golden coins do they have now?

Part – Part- Whole – whole unknown (23/10, 37/10, 20/23)
Mathtastic Level 3 - Numbers to 50
Add and subtract 10
www.tracyashbridge.com

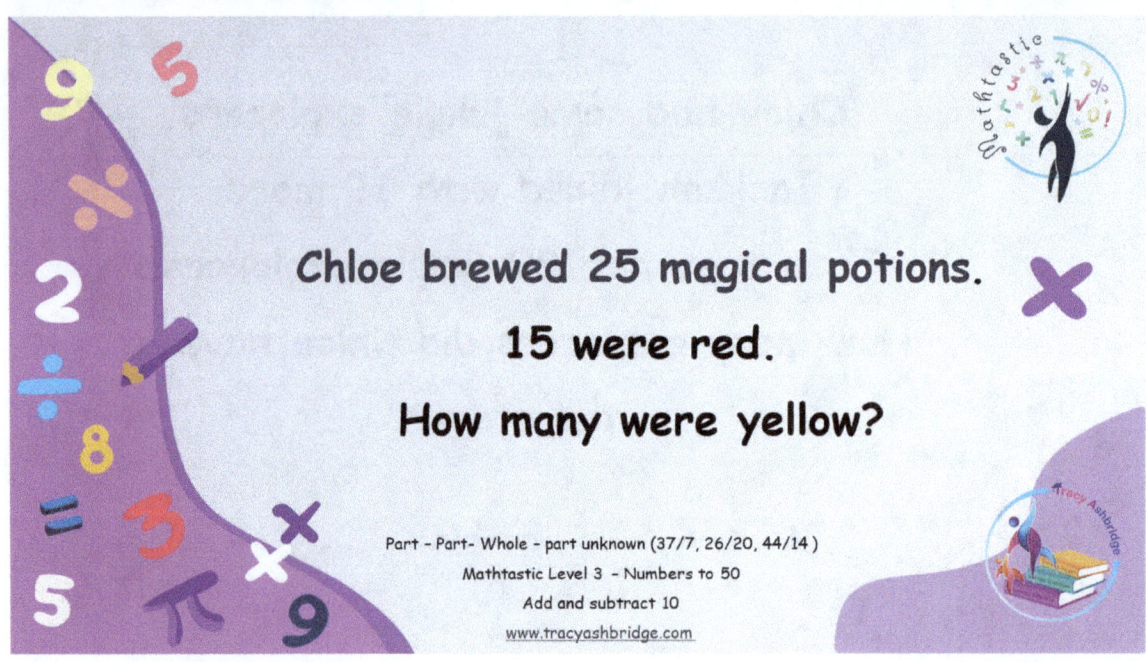

Chloe brewed 25 magical potions.
15 were red.
How many were yellow?

Part – Part- Whole – part unknown (37/7, 26/20, 44/14)
Mathtastic Level 3 - Numbers to 50
Add and subtract 10
www.tracyashbridge.com

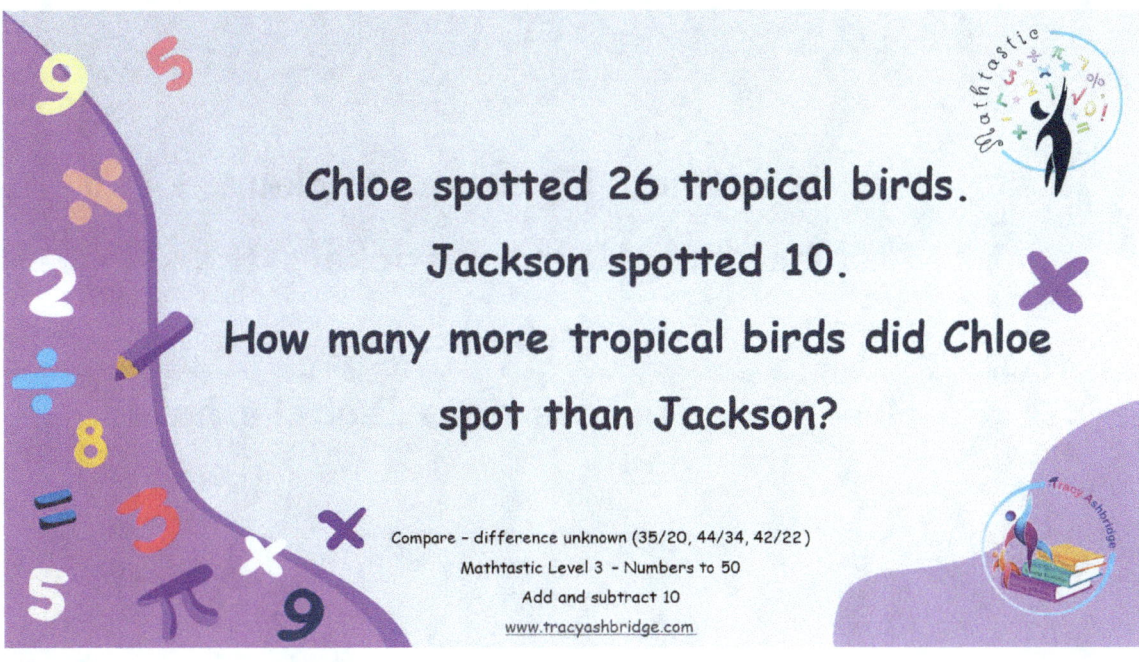

Chloe spotted 26 tropical birds.
Jackson spotted 10.
How many more tropical birds did Chloe spot than Jackson?

Compare – difference unknown (35/20, 44/34, 42/22)
Mathtastic Level 3 – Numbers to 50
Add and subtract 10
www.tracyashbridge.com

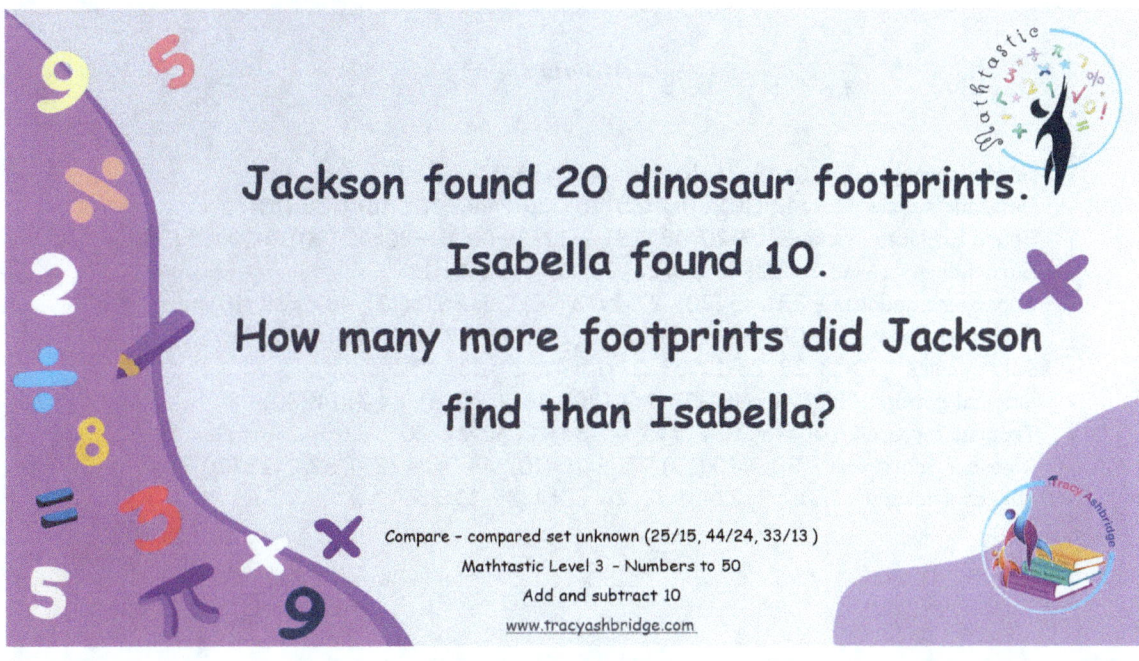

Jackson found 20 dinosaur footprints.
Isabella found 10.
How many more footprints did Jackson find than Isabella?

Compare – compared set unknown (25/15, 44/24, 33/13)
Mathtastic Level 3 – Numbers to 50
Add and subtract 10
www.tracyashbridge.com

© Copyright 2024 Mathtastic: Tracy Ashbridge. All rights reserved

Noah had 17 rocket stickers.
Noah has 10 more stickers than Isabella.
How many stickers does Isabella have?

Compare – referent unknown (22/10, 24/20, 45/10)
Mathtastic Level 3 – Numbers to 50
Add and subtract 10
www.tracyashbridge.com

Answers

- Safari animals – 39+10=49, 26+10=36, 16+20=36, 17+20=37
- Dinosaur eggs – 32+?=42 (10), 10+?=25 (15), 26+?=46 (20), 18+?=28 (10)
- Space explorer stickers – ?+20=48 (28), ?+10=36 (26), ?+20=40 (20), ?+10=45 (35)
- Safari animals – 32-10=22, 46-20=26, 28-20=8, 22-10=12
- Superhero gadgets – 25-?=5 (20), 27-?=17 (10), 33-?=20 (13), 48-?=38 (10)
- Jungle explorers – ?+19=29 (10), ?+20=45 (25), ?+3=37 (30), ?+10=44 (34)
- Golden coins – 10+18=28, 23+10=33, 37+10=47, 20+23= 43
- Magical potions – 25-?=15 (10), 37-?=7 (30), 26-?=20 (6), 44-?=14 (30)
- Tropical birds – 26-10=16, 35-20=15, 44-34=10, 42-22=20
- Dinosaur footprints – 20-?=10 (10), 25-?=15 (10), 44-?=24 (20), 33-?=13 (20)
- Rocket stickers – 17-10=7, 22-10=12, 24-20=4, 45-10=35

Module 5 - Doubles

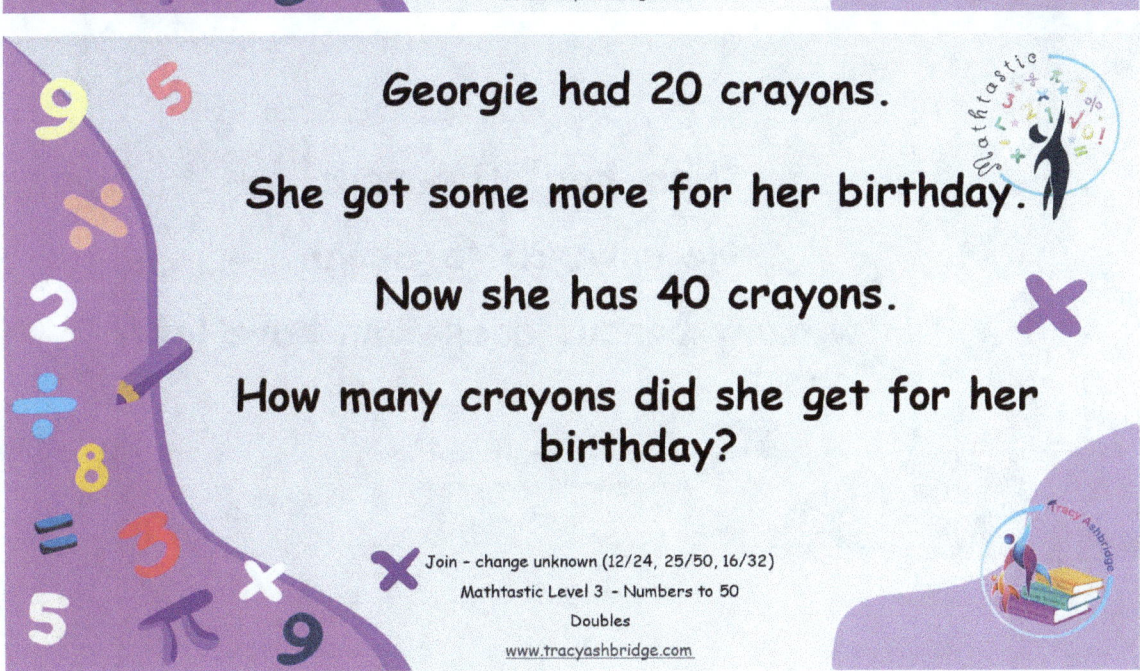

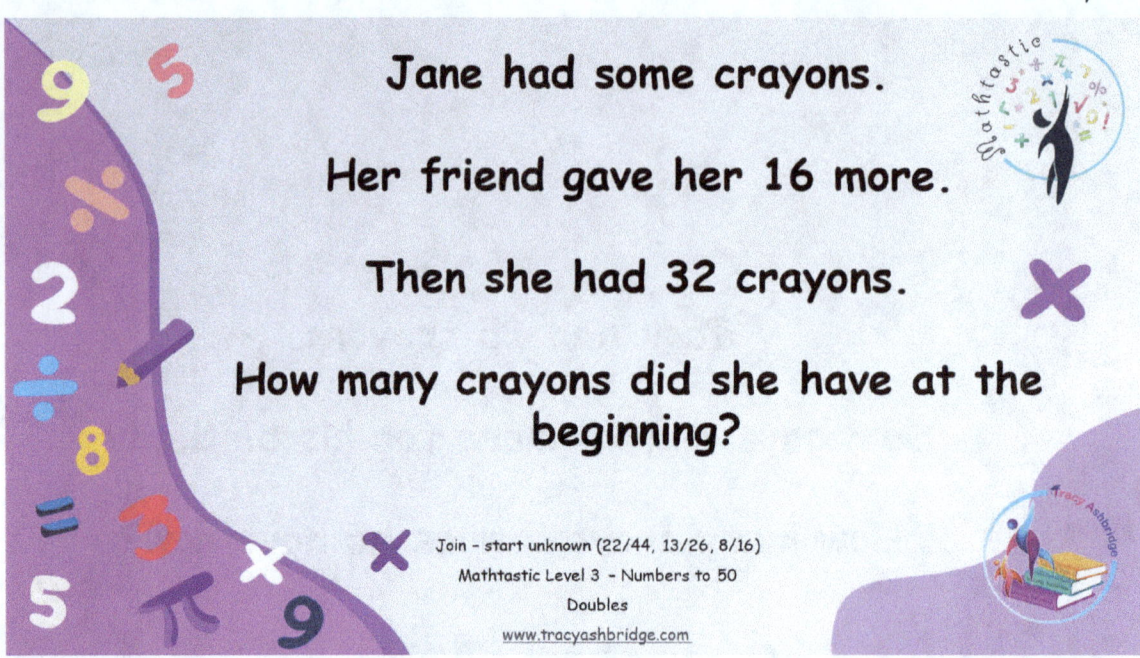

Jane had some crayons.

Her friend gave her 16 more.

Then she had 32 crayons.

How many crayons did she have at the beginning?

Join – start unknown (22/44, 13/26, 8/16)
Mathtastic Level 3 – Numbers to 50
Doubles
www.tracyashbridge.com

Sam had 50 pencils.

He gave 25 to Sarah.

How many pencils does Sam have left?

Separate – result unknown (38/19, 26/13, 16/8)
Mathtastic Level 3 – Numbers to 50
Doubles
www.tracyashbridge.com

© Copyright 2024 Mathtastic: Tracy Ashbridge. All rights reserved

Dave had 38 crayons.
He gave some to Tracy.
Dave has 19 crayons left?
How many did he give to Tracy?

Separate – change unknown (48/24, 34/17, 22/11)
Mathtastic Level 3 – Numbers to 50
Doubles
www.tracyashbridge.com

Tim had some crayons.
He gave 12 to Flynn.
Then he had 12 left.
How many crayons did Tim have to start with?

Separate – start unknown (25/25. 16/16, 23/23)
Mathtastic Level 3 – Numbers to 50
Doubles
www.tracyashbridge.com

© Copyright 2024 Mathtastic: Tracy Ashbridge. All rights reserved

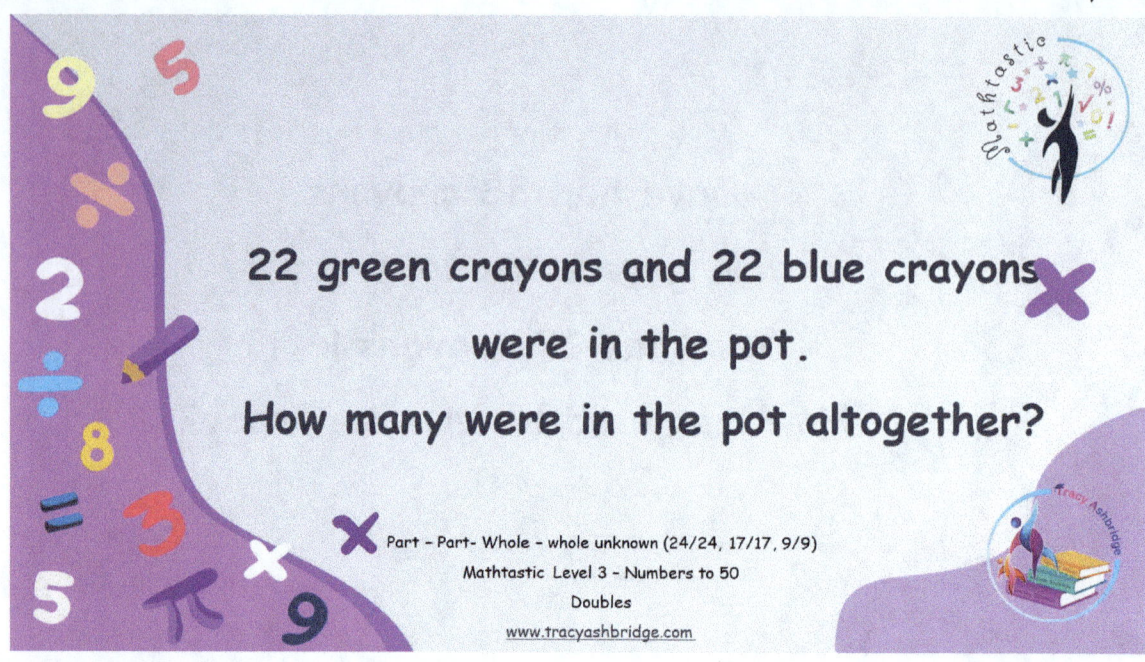

22 green crayons and 22 blue crayons were in the pot.
How many were in the pot altogether?

Part – Part- Whole – whole unknown (24/24, 17/17, 9/9)
Mathtastic Level 3 – Numbers to 50
Doubles
www.tracyashbridge.com

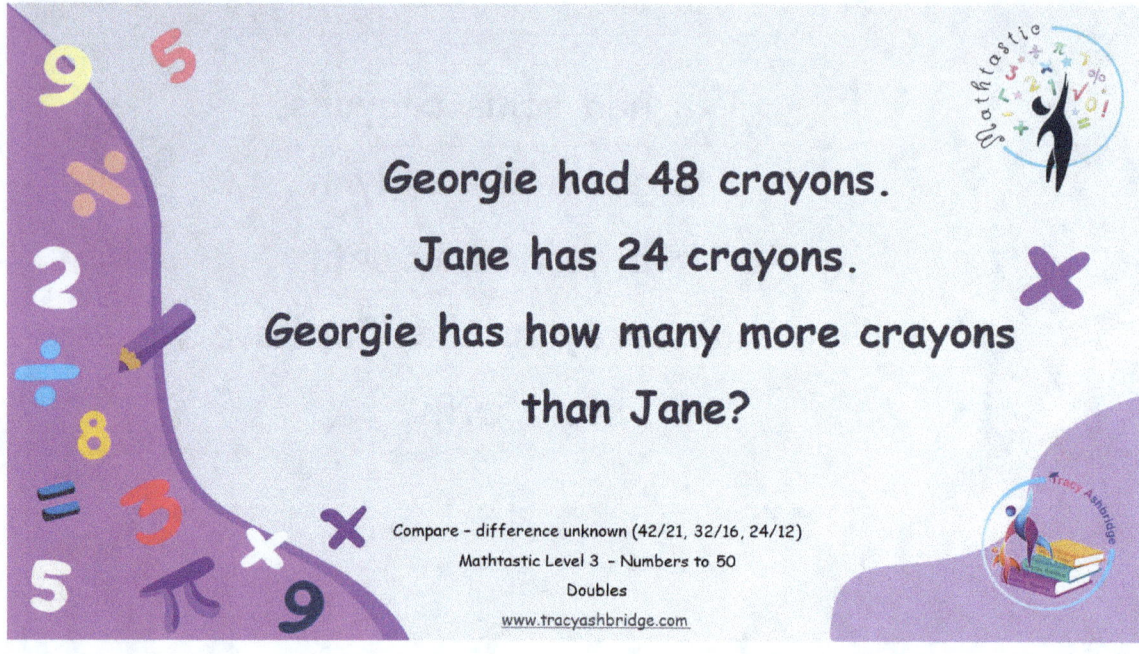

Georgie had 48 crayons.
Jane has 24 crayons.
Georgie has how many more crayons than Jane?

Compare – difference unknown (42/21, 32/16, 24/12)
Mathtastic Level 3 – Numbers to 50
Doubles
www.tracyashbridge.com

© Copyright 2024 Mathtastic: Tracy Ashbridge. All rights reserved

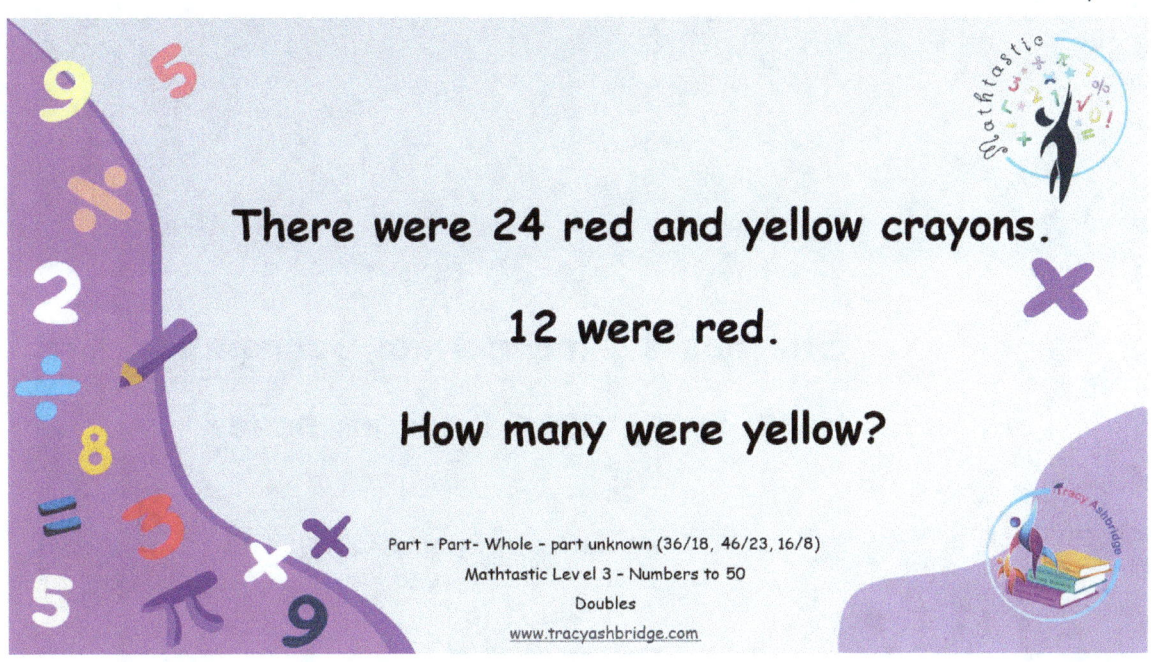

There were 24 red and yellow crayons.

12 were red.

How many were yellow?

Part – Part- Whole – part unknown (36/18, 46/23, 16/8)
Mathtastic Level 3 – Numbers to 50
Doubles
www.tracyashbridge.com

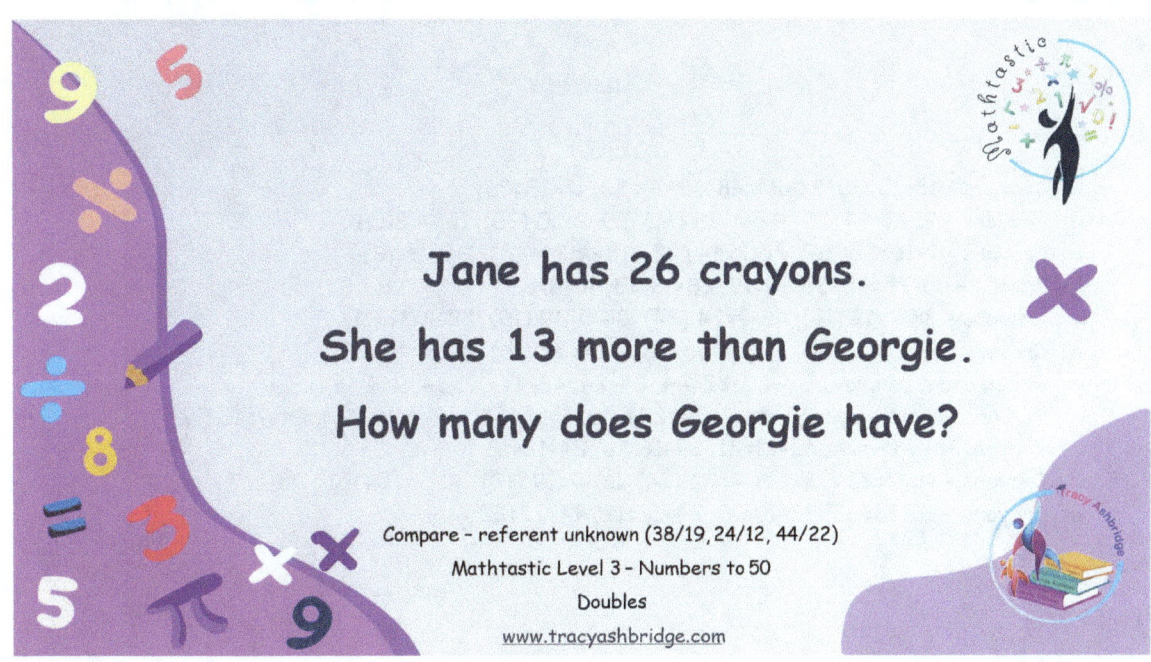

Jane has 26 crayons.

She has 13 more than Georgie.

How many does Georgie have?

Compare – referent unknown (38/19, 24/12, 44/22)
Mathtastic Level 3 – Numbers to 50
Doubles
www.tracyashbridge.com

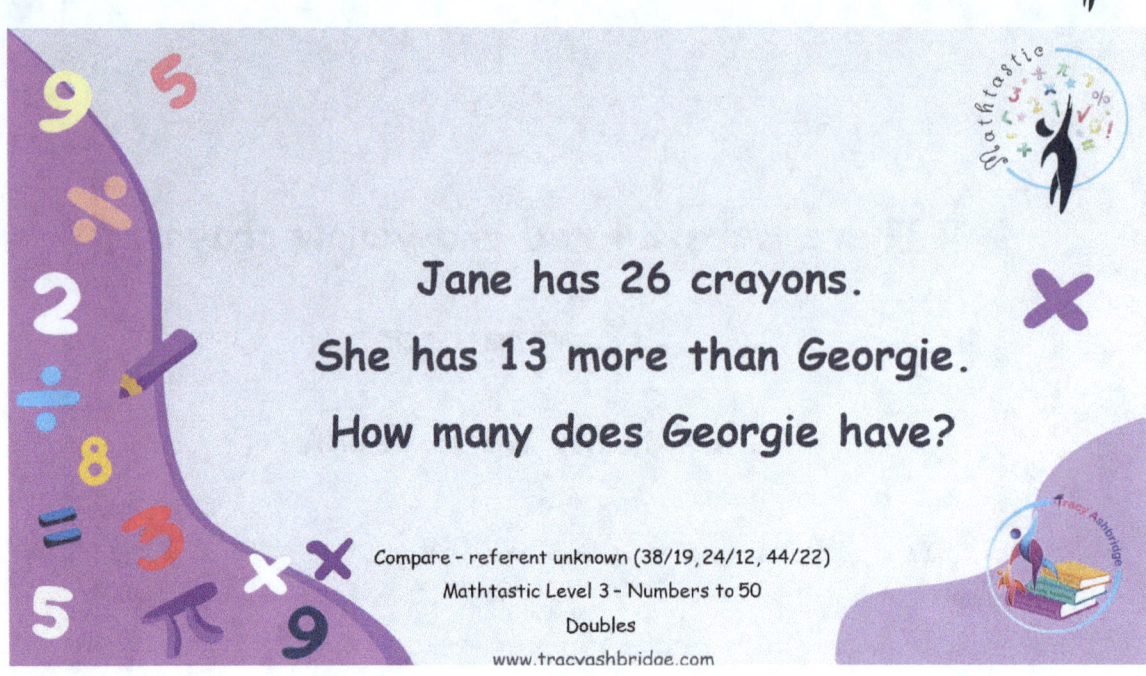

Jane has 26 crayons.

She has 13 more than Georgie.

How many does Georgie have?

Compare - referent unknown (38/19, 24/12, 44/22)
Mathtastic Level 3 – Numbers to 50
Doubles
www.tracyashbridge.com

Answers

- Crayons – 15+15=30, 24+24=48, 19+19=38, 13+13=26
- Crayons – 20+?=40 (20), 12+?=24 (12), 25+?=50 (25), 16+?=32 (16)
- Crayons – ?+16=32 (16), 22+?=44 (22), 13+?=26 (13), ?+8=16 (8)
- Pencils – 50-25=25, 38-19=19, 26-13=13, 16-8=8
- Crayons – 38-?=19 (19), 48-?=24 (24), 34-?=19 (19), 22-?=11 (11)
- Crayons – ?-12=12 (24), ?-25=50 (25), ?-16=32 (16), ?-23=46 (23)
- Green and blue crayons – 22+22=44, 24+24=48, 17+17=34, 9+9=18
- Red and yellow crayons – 48-?=24 (24), 42-?=21 (21), 32-?=16 (16), 24-?=12 (12)
- Crayons – 48-24=24, 42-21=21, 32-16=16, 24-12=12
- Crayons – 46-?=23 (23), 48-?=24 (24), 36-?=19 (19), 30-?=15 (15)
- Crayons – 26-13= 13, 38-19=19, 24-12=12, 44-22=22

Module 6- Near Doubles

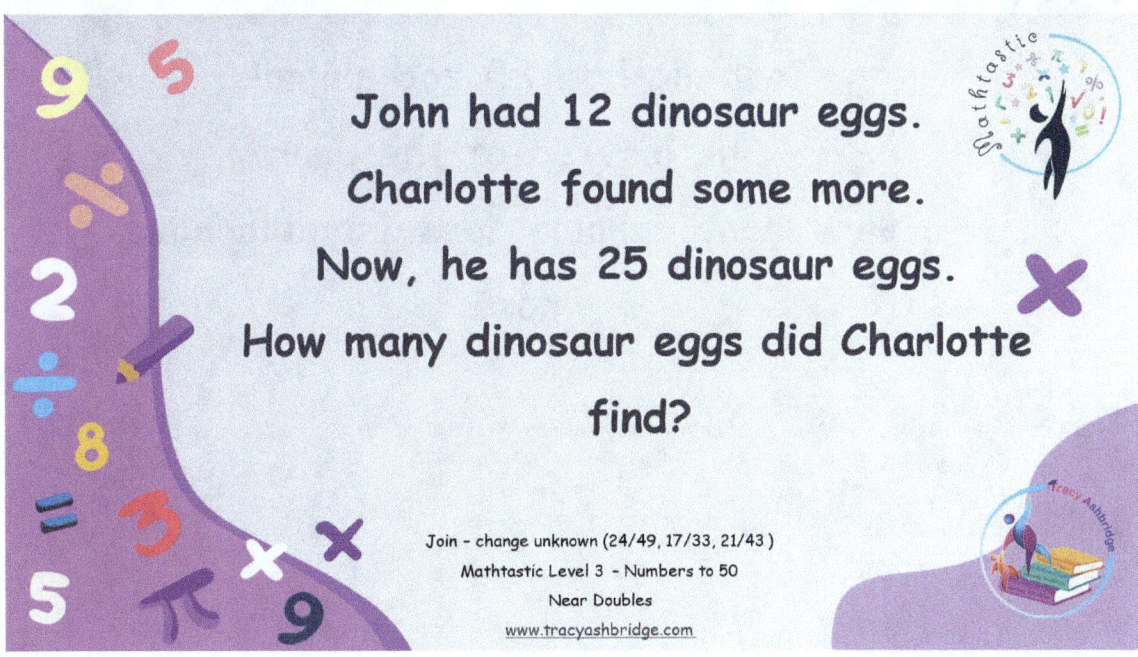

Matt had some space explorer stickers.
Sam gave him 20 more.
Now, he has 41 space explorer stickers.
How many stickers did Sam give him?

Join – start unknown (13/27, 16/31, 22/43)
Mathtastic Level 3 – Numbers to 50
Near Doubles
www.tracyashbridge.com

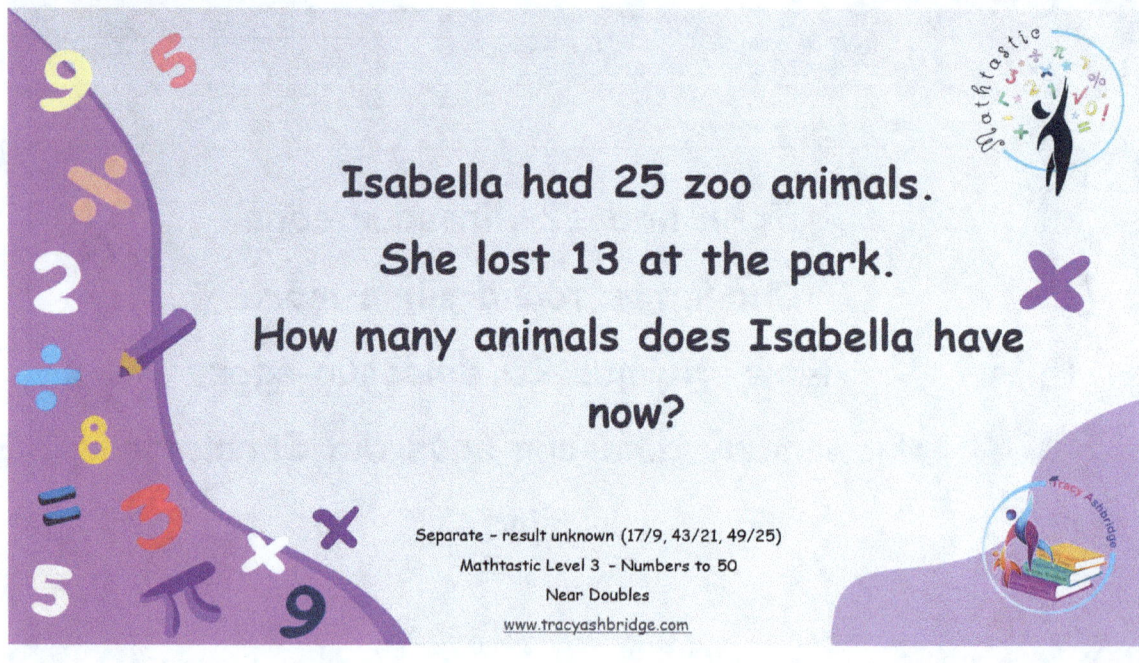

Isabella had 25 zoo animals.
She lost 13 at the park.
How many animals does Isabella have now?

Separate – result unknown (17/9, 43/21, 49/25)
Mathtastic Level 3 – Numbers to 50
Near Doubles
www.tracyashbridge.com

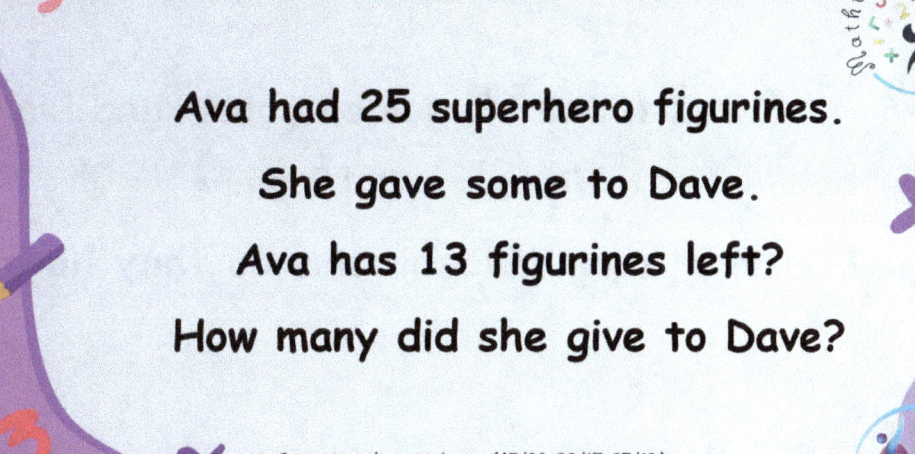

Ava had 25 superhero figurines.
She gave some to Dave.
Ava has 13 figurines left?
How many did she give to Dave?

Separate - change unknown (45/22, 33/17, 27/13)
Mathtastic Level 3 - Numbers to 50
Near Doubles
www.tracyashbridge.com

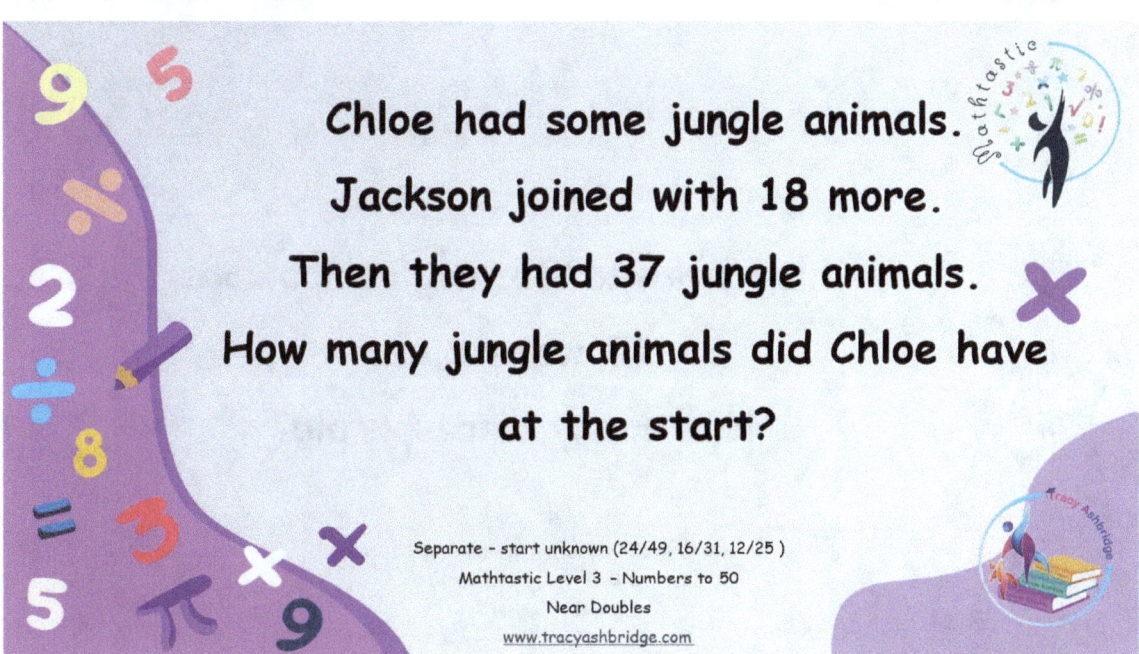

Chloe had some jungle animals.
Jackson joined with 18 more.
Then they had 37 jungle animals.
How many jungle animals did Chloe have at the start?

Separate - start unknown (24/49, 16/31, 12/25)
Mathtastic Level 3 - Numbers to 50
Near Doubles
www.tracyashbridge.com

© Copyright 2024 Mathtastic: Tracy Ashbridge. All rights reserved

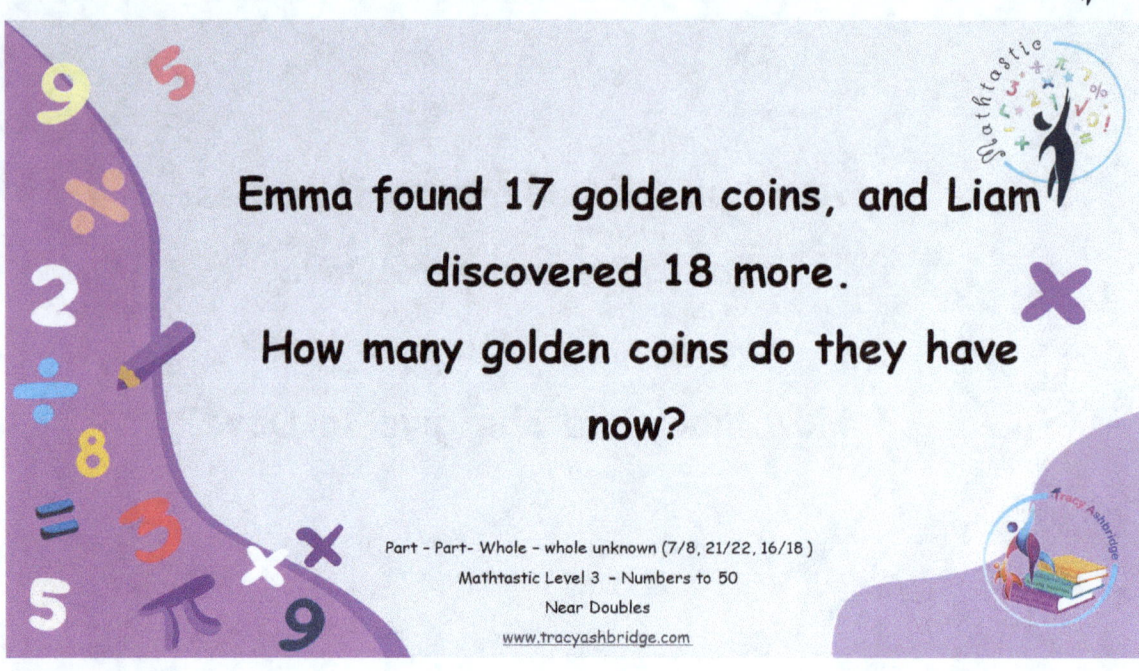

Emma found 17 golden coins, and Liam discovered 18 more.
How many golden coins do they have now?

Part – Part- Whole – whole unknown (7/8, 21/22, 16/18)
Mathtastic Level 3 – Numbers to 50
Near Doubles
www.tracyashbridge.com

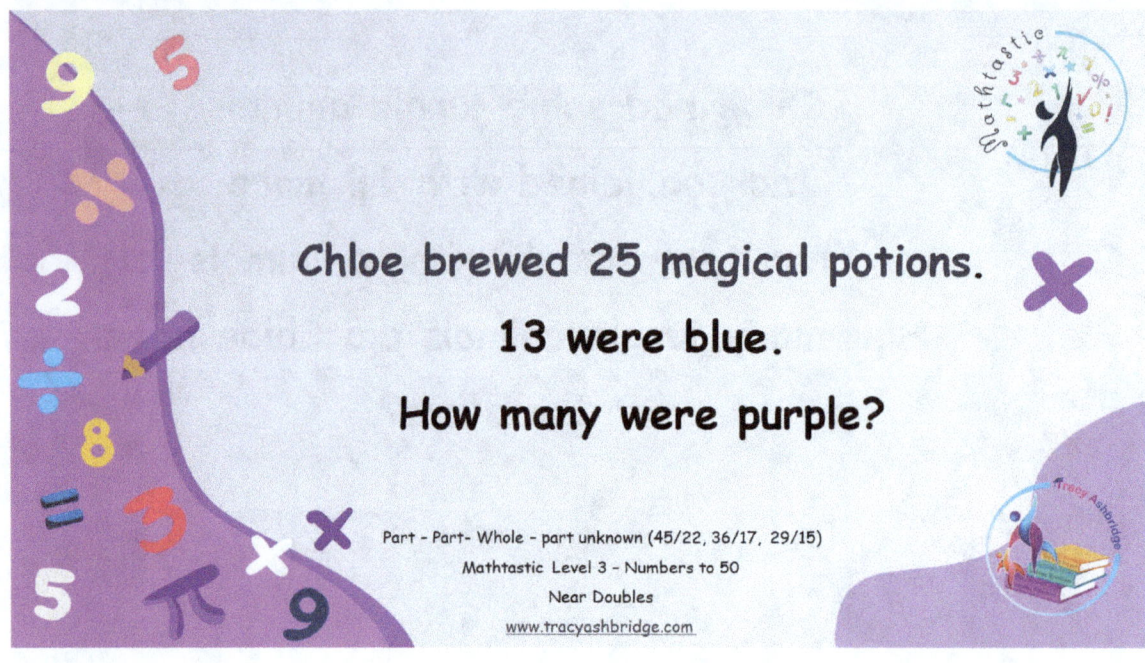

Chloe brewed 25 magical potions.
13 were blue.
How many were purple?

Part – Part- Whole – part unknown (45/22, 36/17, 29/15)
Mathtastic Level 3 – Numbers to 50
Near Doubles
www.tracyashbridge.com

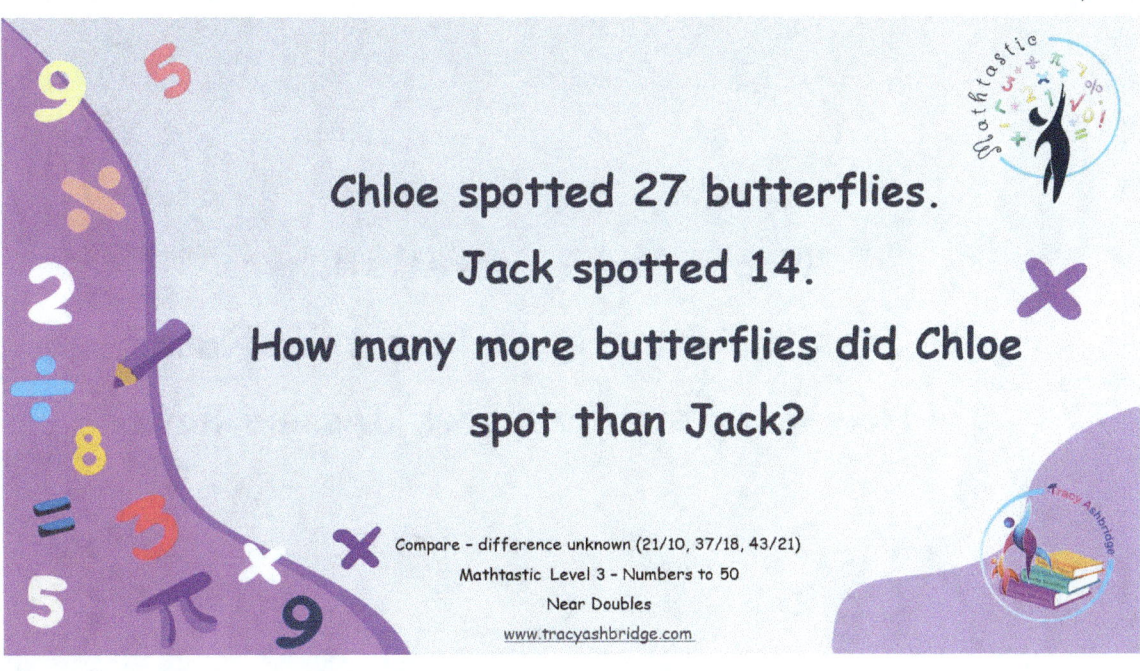

Chloe spotted 27 butterflies.
Jack spotted 14.
How many more butterflies did Chloe spot than Jack?

Compare – difference unknown (21/10, 37/18, 43/21)
Mathtastic Level 3 – Numbers to 50
Near Doubles
www.tracyashbridge.com

Jack found 21 dinosaur footprints.
Bella found 10.
How many more footprints did Jack find than Bella?

Compare – compared set unknown (23/12, 29/15, 45/22)
Mathtastic Level 3 – Numbers to 50
Near Doubles
www.tracyashbridge.com

© Copyright 2024 Mathtastic: Tracy Ashbridge. All rights reserved

Naomi has 9 rocket stickers.
Jess has 8 more stickers than Naomi.
How many stickers does Jessica have?

Compare – referent unknown (13/12, 21/20, 16/18)
Mathtastic Level 3 – Numbers to 50
Near Doubles
www.tracyashbridge.com

Answers

- Jungle animals – 23+22=25, 12+13=25, 17+18=35, 23+35=48
- Dinosaur eggs – 12+?=25 (13), 24+?=49 (25), 17+?=33 (16), 21+?=43 (22)
- Space explorer stickers - ?+20=41 (21), ?+ 13=27 (14), ?+16=31 (15), ?+22=45 (23)
- Zoo animals – 25-?=13 (12), 17-?=9 (8), 43-?=21 (22), 49-?=25 (24)
- Superheroes – 25-?=13 (12), 45-?=22 (23), 33 -?=17 (16), 27-?=13 (14)
- Jungle animals - ?+18=37 (19), 24+?=49 (25), ?+16=31 (15), ?+12=25 (13)
- Golden coins – 17+18=35, 7+8=15, 21+22=43, 16+18= 34
- Magical potions – 25-?=13 (12), 45-?=22 (23), 36 -?=17 (19), 29-?=15 (14)
- Butterflies – 27-14=13, 21-10=11, 37-18=19, 43-21=22
- Dinosaur footprints – 21-?=10 (11), 32-?=12 (11), 29-?=14 (15), 45-?=22 (23)
- Rocket stickers – 8+9=17, 13+12=25, 21+20=41, 16+18=34

© Copyright 2024 Mathtastic: Tracy Ashbridge. All rights reserved

Module 7 – Partition by place value

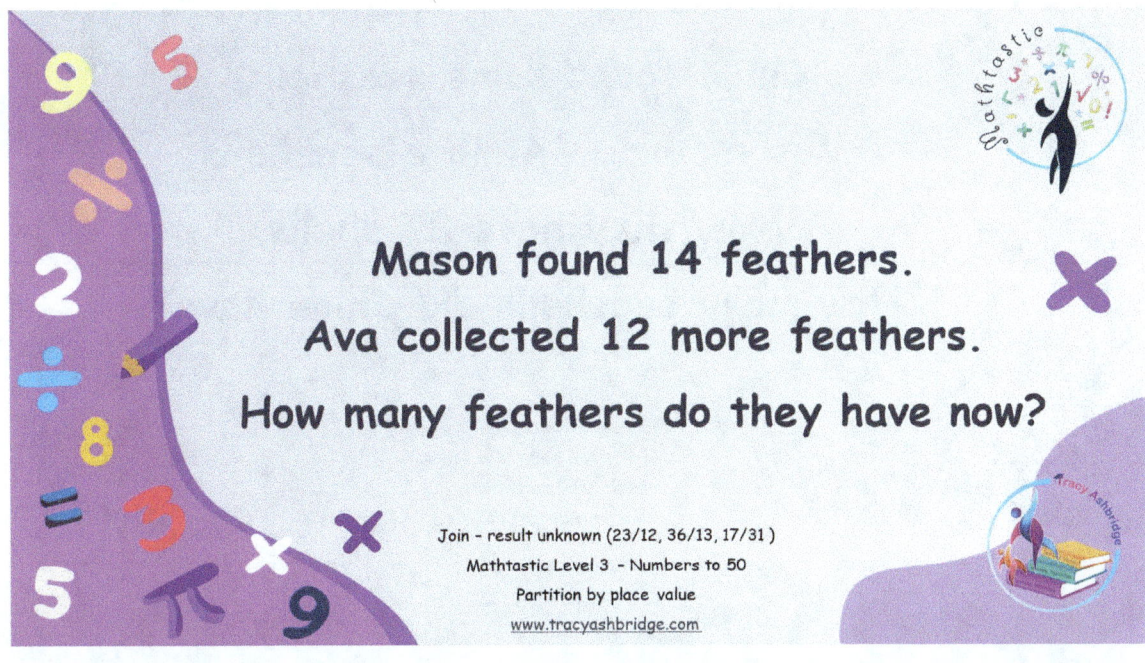

Chloe found some seashells.
Mason found 14 more seashells.
Now they have 27 shells.
How many seashells did Chloe have?

Join – start unknown (15/28, 12/33, 26/48)
Mathtastic Level 3 – Numbers to 50
Partition by place value
www.tracyashbridge.com

Emma spotted 36 monkeys.
Liam saw 13 monkeys leave.
How many monkeys stayed?

Separate – result unknown (22/16, 25/14, 35/15)
Mathtastic Level 3 – Numbers to 50
Partition by place value
www.tracyashbridge.com

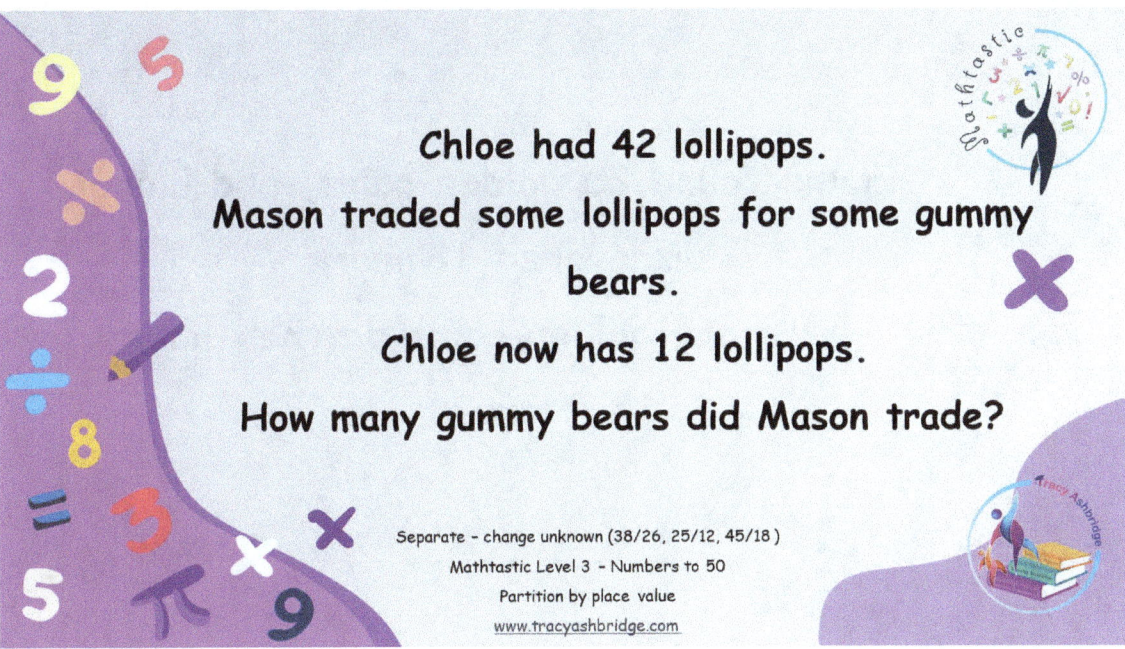

Chloe had 42 lollipops.
Mason traded some lollipops for some gummy bears.
Chloe now has 12 lollipops.
How many gummy bears did Mason trade?

Separate - change unknown (38/26, 25/12, 45/18)
Mathtastic Level 3 - Numbers to 50
Partition by place value
www.tracyashbridge.com

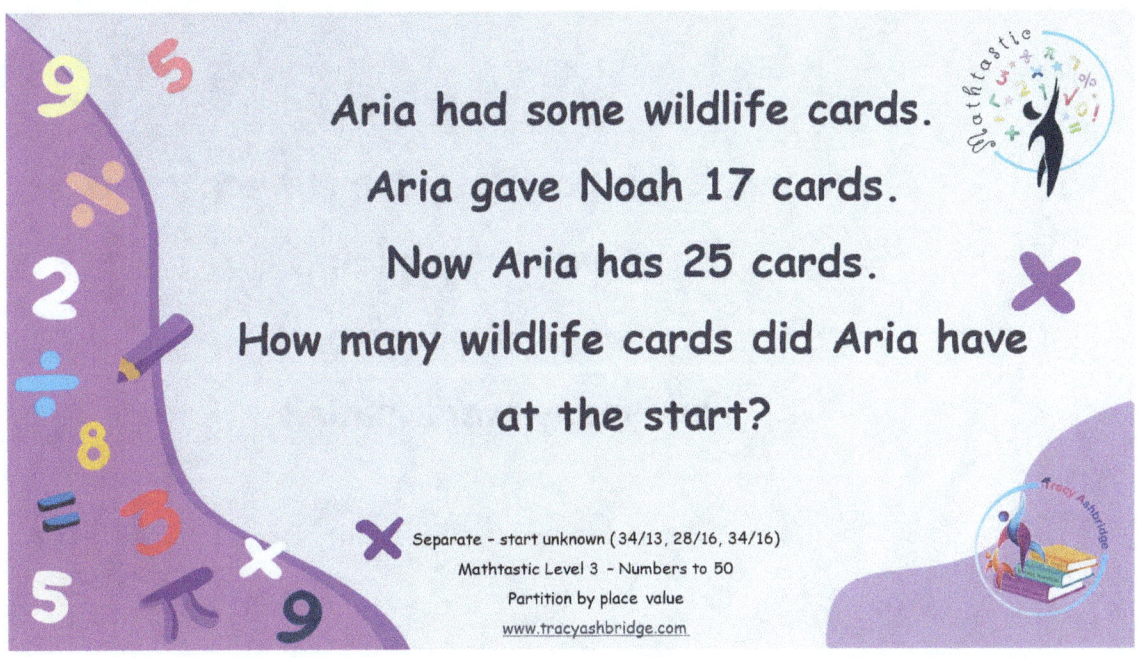

Aria had some wildlife cards.
Aria gave Noah 17 cards.
Now Aria has 25 cards.
How many wildlife cards did Aria have at the start?

Separate - start unknown (34/13, 28/16, 34/16)
Mathtastic Level 3 - Numbers to 50
Partition by place value
www.tracyashbridge.com

Emma found 25 golden coins, and Liam discovered 18 more.
How many golden coins do they have now?

Part – Part- Whole – whole unknown (34/13, 26/18, 14/25)
Mathtastic Level 3 – Numbers to 50
Partition by place value
www.tracyashbridge.com

Chloe made 25 cookies, some were choc chip and some plain.
14 were choc chip.
How many were plain?

Part – Part- Whole – part unknown (46/18, 45/22, 32/15)
Mathtastic Level 3 – Numbers to 50
Partition by place value
www.tracyashbridge.com

© Copyright 2024 Mathtastic: Tracy Ashbridge. All rights reserved

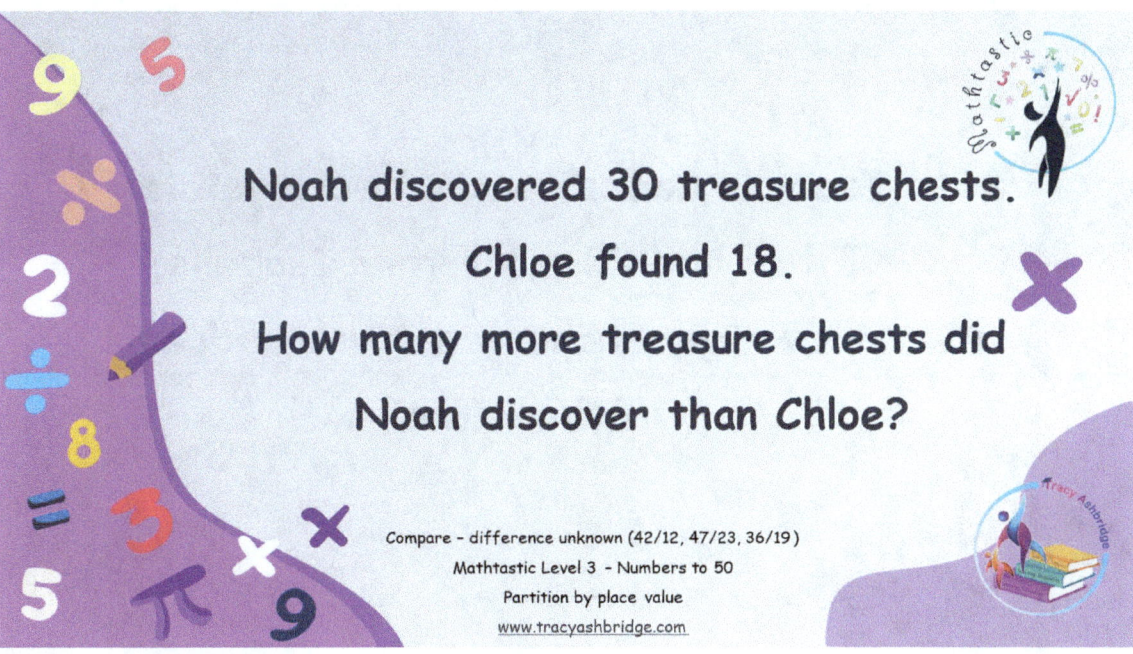

Noah discovered 30 treasure chests. Chloe found 18. How many more treasure chests did Noah discover than Chloe?

Compare - difference unknown (42/12, 47/23, 36/19)
Mathtastic Level 3 - Numbers to 50
Partition by place value
www.tracyashbridge.com

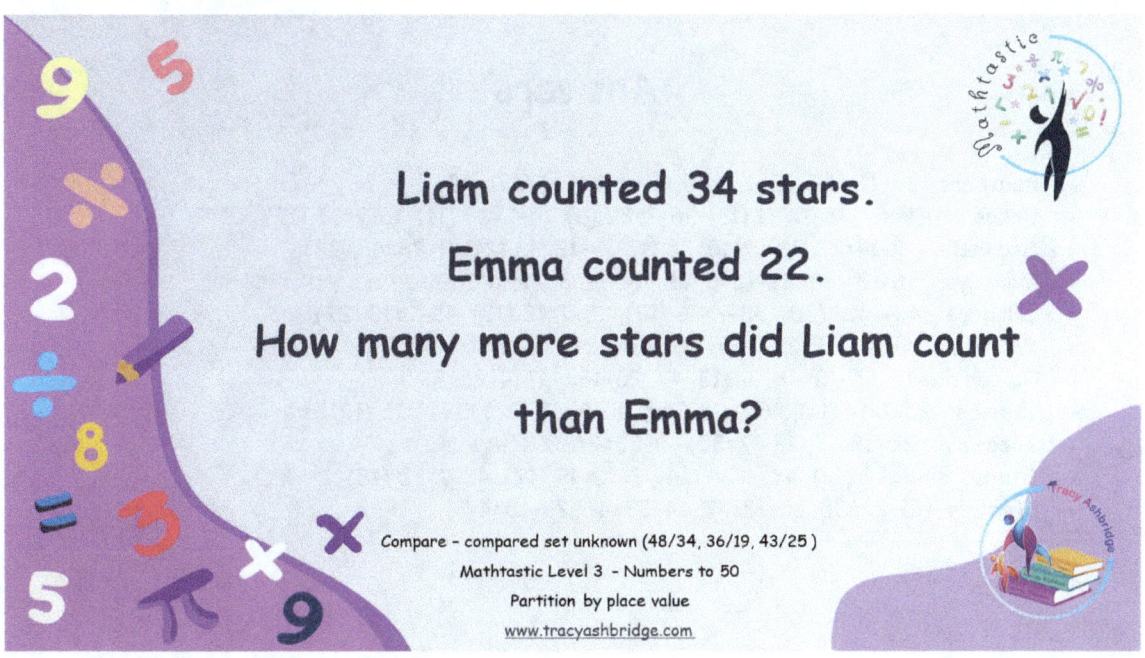

Liam counted 34 stars. Emma counted 22. How many more stars did Liam count than Emma?

Compare - compared set unknown (48/34, 36/19, 43/25)
Mathtastic Level 3 - Numbers to 50
Partition by place value
www.tracyashbridge.com

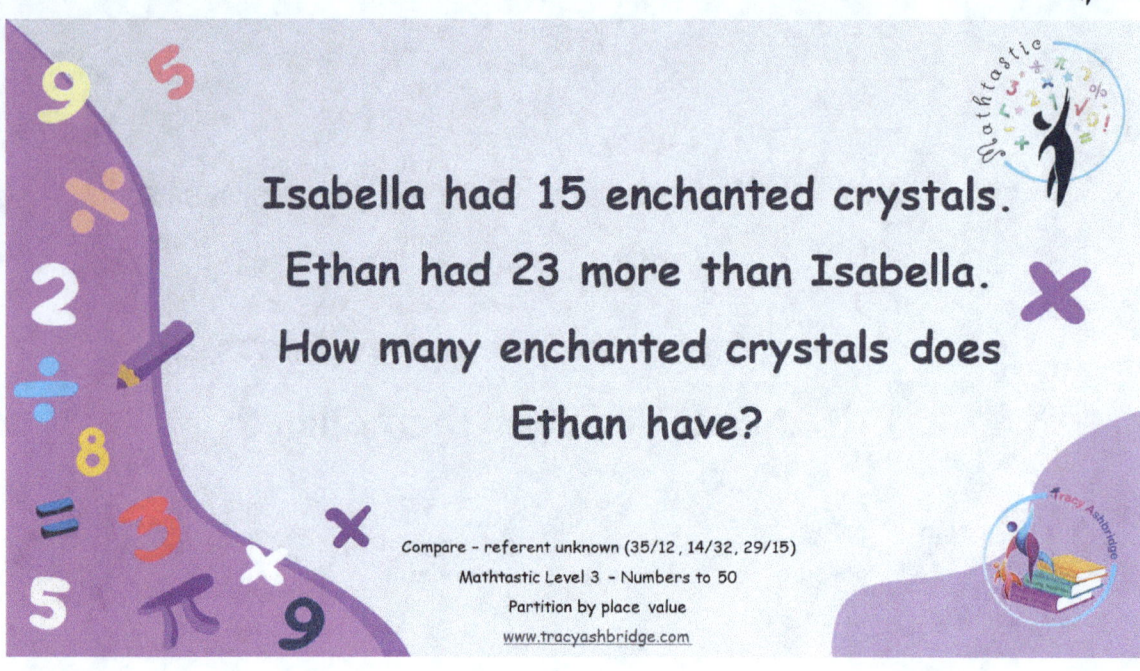

Isabella had 15 enchanted crystals. Ethan had 23 more than Isabella. How many enchanted crystals does Ethan have?

Compare - referent unknown (35/12 , 14/32, 29/15)
Mathtastic Level 3 - Numbers to 50
Partition by place value
www.tracyashbridge.com

Answers

- Feathers - 14+12= 26, 23+12= 35, 36+13=49, 17+31=48
- Dragon stickers - 18+?= 31 (13), 16+?=22 (6), 14+?=25 (11), 16+?=33 (17)
- Seashells - ?+14=27 (13), ?+15=28 (13), ?+12=33 (21), ?+26=48 (22)
- Monkeys – 36-13= 23, 22-16=6, 25 -14=11; 35-15=20
- Lollipops – 42-?=12 (30), 38 -?=26 (12), 25 -?=12 (13), 45-?=18 (27)
- Wildlife cards - ?-17=25 (42), ?-13=34 (21), ?-28=16 (44), ?-34=16 (50)
- Golden coins – 25+18=46, 34+13=47, 26+18=44, 14+25=39
- Cookies – 25-?=14 (11), 46 -?=18 (28), 45-?=22 (22), 32 -?=15 (17)
- Treasure – 30-18=12, 42-12=30, 47-23=24, 36-19=1
- Stars - 34-?=22 (12), 48-?=34 (14), 36-?=19 (17), 43-?=25 (18)
- Crystals – 15+23=38, 35+12=47, 14+32=46, 29+15=44

Module 8 – Add and Subtract by compensation

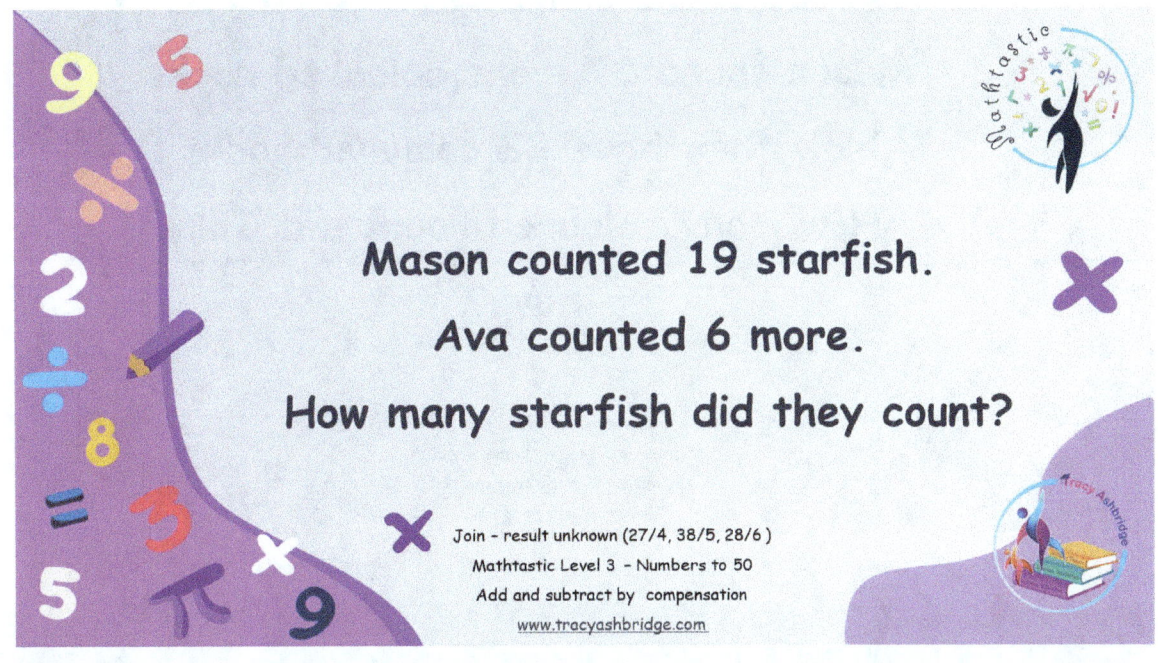

Mason counted 19 starfish.
Ava counted 6 more.
How many starfish did they count?

Join – result unknown (27/4, 38/5, 28/6)
Mathtastic Level 3 – Numbers to 50
Add and subtract by compensation
www.tracyashbridge.com

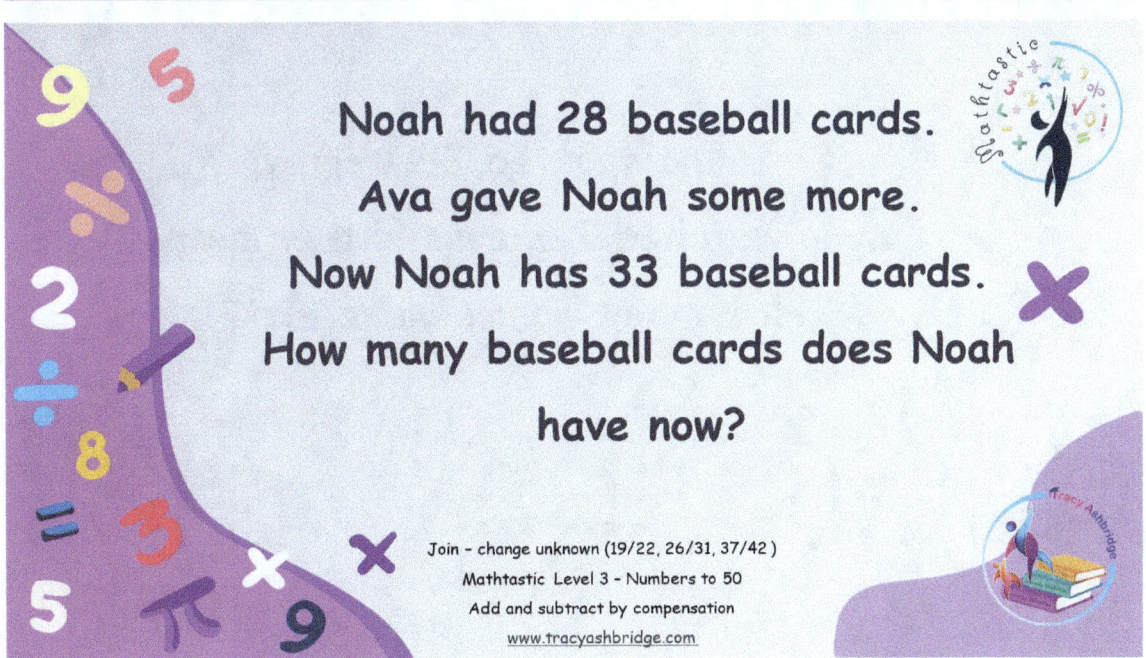

Noah had 28 baseball cards.
Ava gave Noah some more.
Now Noah has 33 baseball cards.
How many baseball cards does Noah have now?

Join – change unknown (19/22, 26/31, 37/42)
Mathtastic Level 3 – Numbers to 50
Add and subtract by compensation
www.tracyashbridge.com

Tillie found some coloured rocks.
Mason found 17 more coloured rocks.
Now they have 22 coloured rocks.
How many coloured rocks did Tillie have?

Join - start unknown (29/35, 6/36, 6/15)
Mathtastic Level 3 - Numbers to 50
Add and subtract by compensation
www.tracyashbridge.com

Emma spotted 36 worms.
Lynn watched 7 worms slither away.
How many worms were left?

Separate - result unknown (42/5, 31/4, 24/5)
Mathtastic Level 3 - Numbers to 50
Add and subtract by compensation
www.tracyashbridge.com

© Copyright 2024 Mathtastic: Tracy Ashbridge. All rights reserved

Zoe had 33 gummy bears.
Fred ate some of the gummy bears.
Zoe now has 27 gummy bears.
How many gummy bears did Fred eat?

Separate - change unknown (32/26, 25/17, 42/38)
Mathtastic Level 3 - Numbers to 50
Add and subtract by compensation
www.tracyashbridge.com

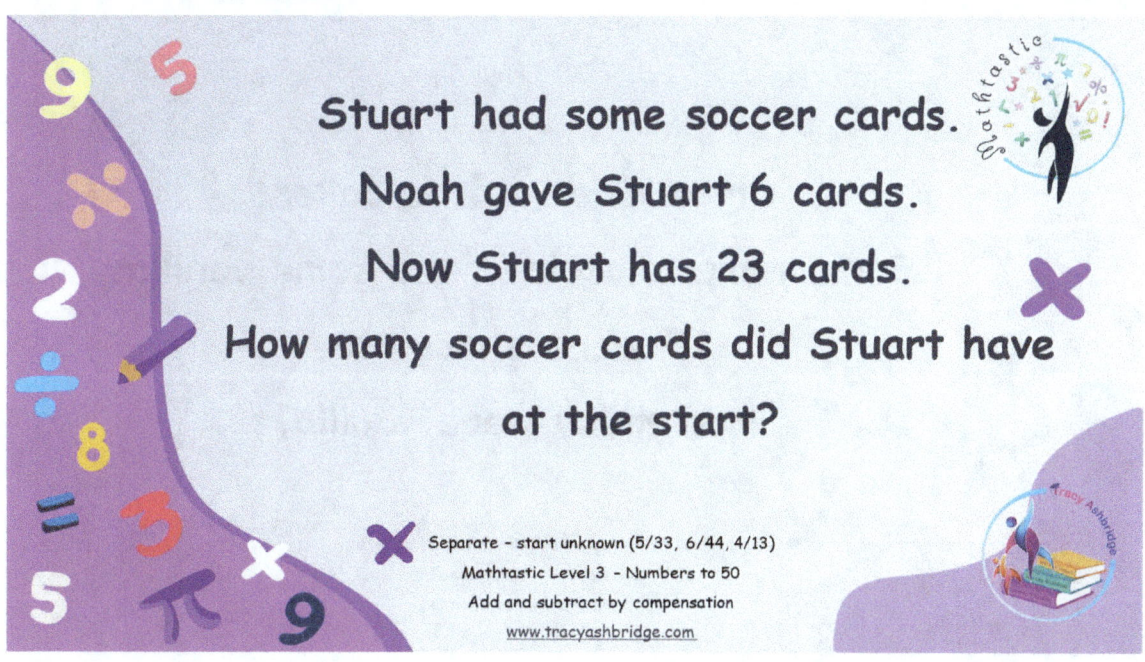

Stuart had some soccer cards.
Noah gave Stuart 6 cards.
Now Stuart has 23 cards.
How many soccer cards did Stuart have at the start?

Separate - start unknown (5/33, 6/44, 4/13)
Mathtastic Level 3 - Numbers to 50
Add and subtract by compensation
www.tracyashbridge.com

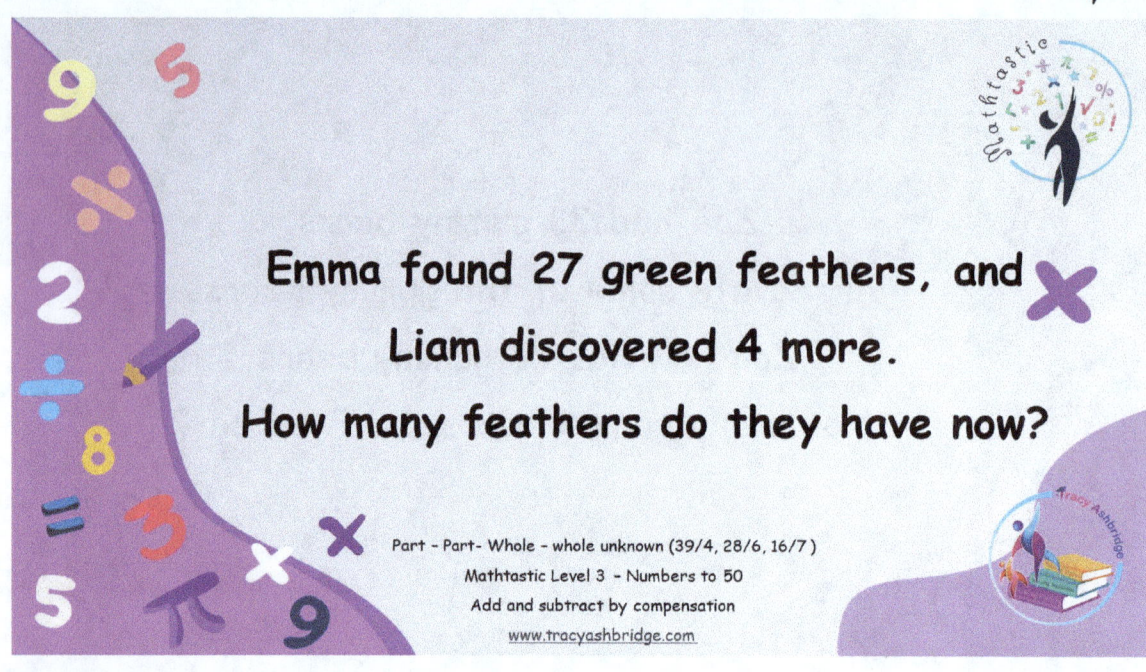

Emma found 27 green feathers, and Liam discovered 4 more.
How many feathers do they have now?

Part – Part- Whole – whole unknown (39/4, 28/6, 16/7)
Mathtastic Level 3 – Numbers to 50
Add and subtract by compensation
www.tracyashbridge.com

Chloe made 25 cupcakes.
Some were chocolate and some vanilla.
19 were chocolate.
How many were vanilla?

Part – Part- Whole – part unknown (36/5, 23/6, 19/5)
Mathtastic Level 3 – Numbers to 50
Add and subtract by compensation
www.tracyashbridge.com

Liam counted 27 shooting stars. Emma observed 6. How many more shooting stars did Liam see than Emma?

Compare - difference unknown (34/7, 43/4, 30/5)
Mathtastic Level 3 - Numbers to 50
Add and subtract by compensation
www.tracyashbridge.com

Mason picked 26 ripe apples. Olivia picked 6 more than Mason. How many apples did Olivia pick?

Compare - compared set unknown (37/5, 16/6, 29/4)
Mathtastic Level 3 - Numbers to 50
Add and subtract by compensation
www.tracyashbridge.com

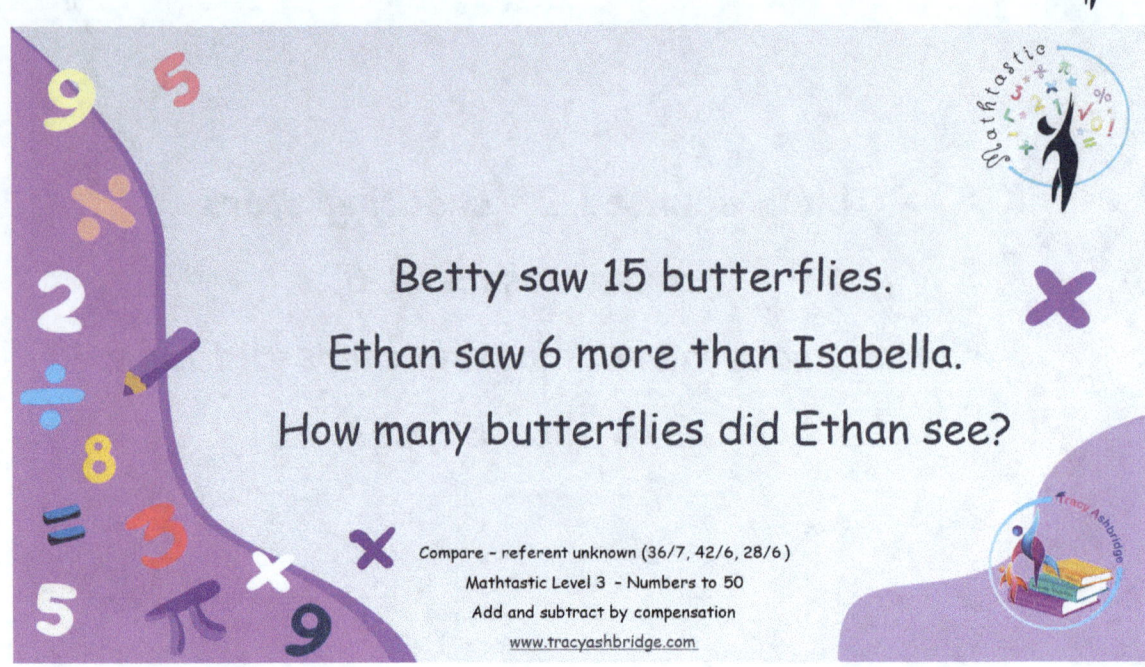

Betty saw 15 butterflies.
Ethan saw 6 more than Isabella.
How many butterflies did Ethan see?

Compare – referent unknown (36/7, 42/6, 28/6)
Mathtastic Level 3 – Numbers to 50
Add and subtract by compensation
www.tracyashbridge.com

Answers

- Starfish – 19+6=15, 27+4=33, 38+5=43, 28+6=34
- Baseball cards – 28+?=33 (5), 19+?=22 (3), 26+?=31 (5), 37+?=42 (5)
- Coloured rocks – ?+17=22 (5), ?+29=35 (6), ?+6=36 (30), ?+6=15 (9)
- Worms – 36-7=29, 42-5=37, 31-4=27, 24-5=19
- Gummy bears – 33-?=27 (6), 32-?=26 (6), 25-?=17 (8), 42-?=38 (4)
- Soccer cards – ?+6=23 (17), ?+5=33 (27), ?+6=44 (38), ?+4=13 (9)
- Feathers – 27+4=31, 39+4=43, 28+6=34, 16+7=23
- Cupcakes – 25-?=19 (6), 36-?=5 (31), 23-?=6 (17), 19-?=5 (14)
- Shooting stars – 27-6=21, 34-7=27, 43-4-39, 30-5=25
- Apples – 26-?=6 (20), 37-?=6 (21), 16-?=6 (10), 29-?=4 (23)
- Butterflies – 15+6=21, 36+7=43, 42+6=48, 28+6=34

www.ingramcontent.com/pod-product-compliance
Lightning Source LLC
Chambersburg PA
CBHW081418300426
44109CB00019BA/2336